Tourism Transformations in Protected Area Gateway Communities

Tourism Transformations in Protected Area Gateway Communities

Edited by

Susan L. Slocum

Peter Wiltshier

John B. Read IV

CABI

CABI is a trading name of CAB International

CABI
Nosworthy Way
Wallingford
Oxfordshire OX10 8DE
UK

Tel: +44 (0)1491 832111
E-mail: info@cabi.org
Website: www.cabi.org

CABI
WeWork
One Lincoln St
24th Floor
Boston, MA 02111
USA

T: +1 (617)682-9015
E-mail: cabi-nao@cabi.org

A catalogue record for this book is available from the British Library, London, UK.

Library of Congress Cataloging-in-Publication Data

Names: Slocum, Susan L., editor. | Wiltshier, Peter, editor. | Read, John B., IV, editor.
Title: Tourism transformations in protected area gateway communities / edited by Susan L. Slocum, Peter Wiltshier, John B. Read IV.
Description: Wallingford, Oxfordshire ; Boston, MA : CABI, 2022. | Includes bibliographical references and index. | Summary: "Gateway communities are those located adjacent to protected areas and are often the communities most impacted by tourism visitation and are dependent on tourism revenue. This book presents informed, interpreted, and nuanced approaches towards protect area management and conservation, based on bottom-up local experiences by the gateway communities"-- Provided by publisher.
Identifiers: LCCN 2021052220 (print) | LCCN 2021052221 (ebook) | ISBN 9781789249033 (hardback) | ISBN 9781789249040 (ebook) | ISBN 9781789249057 (epub)
Subjects: LCSH: Tourism--Environmental aspects. | Ecotourism. | Protected areas--Public use.
Classification: LCC G156.5.E58 T875 2022 (print) | LCC G156.5.E58 (ebook) | DDC 333.72--dc23/eng/20211223
LC record available at https://lccn.loc.gov/2021052220
LC ebook record available at https://lccn.loc.gov/2021052221

ISBN-13: 9781789249033 (hardback)
9781789249040 (ePDF)
9781789249057 (ePub)

DOI: 10.1079/9781789249033.0000

Commissioning Editor: Claire Parfitt
Editorial Assistant: Lauren Davies
Production Editor: Tim Kapp

Typeset by Exeter Premedia Services Pvt Ltd, Chennai, India
Printed and bound in the UK by Severn, Gloucester

Dedication

Professor Alan Clarke, 1956–2021

Alan's commitment to protected spaces and gateway communities has been marked throughout his career, beginning with his teaching in community education in Sheffield, followed by time with the community education team at the Open University. He has led research in the inner city of Salford and worked with the Roma community in the North West of England. The Derby Jubilee Community Festival crystallized a number of issues – power, hegemony, inclusion, engagement and capacity building. Partnership working motivated Alan to develop solutions to be used by communities in the Baltic States, Croatia, Hungary and the United Kingdom. Alan moved to Hungary where there were hard lessons in coming to terms with cultures, structures and stakeholder relations that have to be seen through differently sensitized lenses. This opened up an interest in all aspects of stakeholder-rich leisure, tourism and events. Alan firmly contributed to our understanding of the ways that belief systems, narratives and interpretation have an impact on the cultures in which they are experienced in the communities that give them meaning, life and a destiny characterized by hope and great positivity.

Contents

List of Tables

List of Figures

Contributors

Dorothee Bohn, PhD candidate in Human Geography at Umeå University, Umeå, Sweden. Dorothee is a member of the Formas funded research project, Climate Change and the Double Amplifications of Arctic Tourism: Challenges and Potential Solutions for Tourism and Sustainable Development in an Arctic Context. Her dissertation examines the spatial and political dimensions of tourism governance in the European Arctic. In addition to tourism policy and planning, Dorothee is interested in exploring labour relations in the service sector, tourism evaluation and issues surrounding the economy–environment nexus.

Andrea Z. Botelho, postdoc researcher, Environmental Science/Tourism, University of the Azores, Ponta Delgada, Portugal. Andrea has a masters degree in Nature Conservation and Management and a PhD in Marine Sciences (Marine Ecology) from the University of the Azores. She has been developing research activities in the fields of environmental planning and management, tourism, biology and marine ecology at the CIBIO/University of the Azores.

Kelly S. Bricker, Professor and Director of Hainan University/Arizona State University Joint International Tourism College, Arizona State University, Phoenix, Arizona, USA. Kelly has research and teaching interests in nature-based tourism and recreation, sense of place, gateway communities, natural resource management, the health benefits of nature-based experiences and the impacts associated with tourism and recreation in natural areas. She serves on the boards of the Global Sustainable Tourism Council, the Tourism and Protected Area Specialist Group of the IUCN and Tourism Unity.

Robert S. Bristow, Professor, Department of Geography, Planning, and Sustainability, Westfield State University, Massachusetts, USA. Rob's 1975, 2000-mile trek of the Appalachian Trail (AT) inspired a career in natural resource planning and management with roles with the Maryland State Park System, internships with the National Park Service and finally, a doctorate from Southern Illinois University at Carbondale. When not teaching, he is a volunteer boundary corridor and cultural resource monitor for the AT in Massachusetts.

Karina Hernández-Casimiro, Administrative Manager/Professor, Autonomous University of Ciudad Juárez, Ciudad Juárez, Mexico. Karina has participated in various national and international congresses related to tourism, information and communication technologies (ICTs), sustainability and administrative sciences. She is the editor responsible for infomercials for the business magazine *MACH*. Her research lines include innovation, ICTs and competitiveness, smart cities, and smart destinations.

Rosa Herminia Suárez-Chaparro, Professor/researcher, Tourism Bachelor Programme, Autonomous University of Ciudad Juárez, Ciudad Juárez, Mexico. Rosa's research interests include landscape analysis and management in natural spaces, tourism management in protected natural areas and sustainable tourism development. She has experience in research projects mainly in PNA of Chihuahua, Janos Biosphere Reserve, Mapimí Biosphere Reserve, APFF Santa Elena Canyon and APFF Médanos de Samalayuca, and is the Coordinator of the International Network of Tourism and Landscape.

Ana C. Costa, Assistant Professor, Department of Biology, University of the Azores, Ponta Delgada, Portugal. Ana is a marine biologist and gained a PhD in Marine Sciences (Marine Ecology) in 2004. She has been developing teaching and research activities in her department since 1990. Her main interest is aquatic biodiversity and ecology and, to a lesser extent, invertebrate taxonomy. She has an interest in marine environmental monitoring (under WFD and MDFD), conservation and planning (CZMP and MSP). Her more recent research has been in aquatic non-indigenous species. She teaches several courses in the biology undergraduate, masters and PhD programmes and has supervised several grad students.

Kynda R. Curtis, Professor, Department of Applied Economics, Utah State University, Logan, Utah, USA. Kynda is a Professor and Extension Specialist in the Department of Applied Economics at Utah State University. She received her PhD in Economics from Washington State University in 2003. Her research interests include agriculture/food marketing, producer adoption and consumer demand for value-added, specialty and differentiated foods. As an Extension Specialist, Professor Curtis assists small-scale growers in developing new markets for their products and assessing the feasibility of new food products and value-added processes.

Margaret J. Daniels, Professor, Tourism and Events Management, George Mason University, Virginia, USA. Maggie has conducted extensive fieldwork in the areas of tourism planning and policy, wedding planning, park visitation and event management as pertaining to local economic development. She partners with agencies in the Washington DC metropolitan area to assist them with event and tourism implementation and evaluation. She has acted as the lead investigator on collaborative studies with the National Park Service regarding the National Mall and Memorial Parks and with the Maryland-National Capital Park and Planning Commission.

Edieser Dela Santa, Professor, University of the Philippines, Asian Institute of Tourism, Quezon City, Metro Manila, Philippines. Edieser is the former Dean of the University of the Philippines Asian Institute of Tourism. His recent works and publications have examined the politics of tourism policy, the representation of culture and heritage, and tourism and hospitality curriculum development. He is currently involved in research projects that look into the governance aspects of protected area tourism, small islands tourism and tourism education.

C. Michael Hall, Professor, Department of Management, Marketing and Entrepreneurship, University of Canterbury, New Zealand. He is also Visiting Professor and Docent in Geography, University of Oulu, Finland; Guest Professor, Department of Service Management and Service Studies, Lund University, Helsingborg, Sweden; Visiting Professor, Department of Organisation and Entrepreneurship, School of Business and Economics, Linnaeus University, Kalmar, Sweden; and Visiting Professor, CRIC, Taylor's University, Malaysia.. He is the co-editor of *Current Issues in Tourism* and has published widely on tourism, regional development, global environmental change, sustainability, degrowth, behavioural change and demarketing, food and world heritage.

Manuel Ramón Gonzalez Herrera, researcher, Autonomous University of Ciudad Juárez, Ciudad Juárez, Mexico. Manuel is a teacher and researcher on topics such as geography of tourism, landscapes, environment and sustainability, tourism planning and management, tourist destinations, research and scientific publication, in countries such as Spain, Cuba, Peru, Mexico, Chile, Brazil, Nicaragua, Canada and the USA. His research projects are related to sustainable tourism management, and he is a director of postdoctoral studies, doctoral theses and masters theses. He has authored of 16 books, 27 book chapters, 60 articles in scientific journals and 43 articles in event reports and has also collaborated on different editorial committees of scientific journals and events.

Russell M. Hicks, PhD Student, Department of Horticulture and Natural Resources, Kansas State University, Manhattan, Kansas, USA. Russell is a doctoral research assistant and student studying Park Management and Conservation in the Department of Horticulture and Natural Resources at Kansas State University. He is currently researching the impacts of rock climbing in the National Capital Region of the National Park Service.

Julie Judkins, Director of Education and Outreach, Appalachian Trail Conservancy, and is based out of Asheville, North Carolina, USA. Julie leads the greater ecosystem of communities and people to protect, manage and advocate for the Trail through new partnerships and programmes. Julie has been with ATC since 2004. Before ATC, Julie served as a Peace Corps volunteer in Armenia and worked with Hurricane Island Outward Bound in Key Largo, an at-risk youth programme. She has a BA in Communications from North Carolina State University and a Masters of Environmental Management through Duke's Environmental Leadership programme at the Nicholas School of Environment.

N. Qwynne Lackey, Assistant Professor, Recreation, Parks and Leisure Studies Department, State University of New York System, Cortland, New York, USA. Qwynne is a recent graduate from the University of Utah. Her primary research interests include sustainable tourism, park and protected area management and community development. Qwynne is also a member of the National Parks Conservation Association's Next Generation Advisory Council.

Natalya Lawrence, Offshore Islands Conservation Programme Coordinator, Environmental Awareness Group, Antigua and Barbuda. Natalya has worked at the Environmental Awareness Group for ten years, working to conserve endangered, endemic species. She is also a member of BirdsCaribbean, the largest regional environmental NGO. She lecturers in the area of leisure and recreation management, as well as global tourism impacts, and she actively works along with community groups, in addition to various NGOs.

Gustavo C.X.M.P. Machado, Associate Professor at the Interdisciplinary Center for Social Development in the Federal University of Rio de Janeiro (NIDES/UFRJ), Rio de Janeiro, Brazil. Gustavo is a postdoc researcher in the area of Public Health and Environment at the National School of Public Health. He received his PhD in Community Psychosociology and Social Ecology from the Federal University of Rio de Janeiro (EICOS/UFRJ).

Gianna Moscardo, Professor, College of Business, Law and Governance, James Cook University, Townsville, Australia. Gianna's research interests include customer and tourist experience design, heritage interpretation, sustainability communication and new approaches to tourism planning and development for sustainability. She is the current Chair of the Building Excellence in Sustainable Tourism Education Network (BEST EN), an international organization committed to the creation and dissemination of knowledge to support sustainable tourism education, research and practice.

Jake Powell, Extension Specialist and Assistant Professor, Department of Landscape Architecture and Environmental Planning, Utah State University, Logan, Utah, USA. Jake's professional career has focused on uniting communities to their surrounding landscapes through collaborative planning and design efforts. He has worked throughout the Intermountain West to envision and implement watershed planning and restoration projects, natural resource conservation efforts, recreation infrastructure and community economic growth opportunities. Jake's research and extension efforts focus on analysing and improving how the design of communities, sites, infrastructure and natural areas affects the quality of the environment and the human experience from the site to the watershed scale.

Sidnei Raimundo, Associate Professor, School of Arts, Sciences and Humanities, University of São Paulo, São Paulo, Brazil. Sidnei received a PhD in Geography (environmental analysis and territorial dynamics) from the Campinas State University (UNICAMP) and served as a postdoc researcher at the University of Girona, Spain. He has experience in protected areas, urban parks and other outdoors spaces, specifically researching governance and management, the geography of leisure and tourism, natural resource management, ecotourism and the impacts of recreation and tourism.

Mary Anne Ramos-Tumanan, Head of the Department of Hotel, Restaurant and Institution Management, University of the Philippines, Quezon City, Metro Manila, Philippines. Mary Anne earned her research doctorate in Tourism Management from the University of Waikato in New Zealand. She helped establish the Raglan Monitoring Center under the UNWTO-INSTO programme. Her research interests include tourism impacts on place attachment and place meaning. Her current research examines the impacts of tourism on the *Ivatans*, the indigenous peoples of the Batanes Archipelago Islands. This longitudinal study is envisaged to be under the auspices of UNWTO-INSTO in the long run.

John B. Read IV, PhD student, University of Waterloo, Ontario, Canada. John has a background in fire protection with experience working as a firefighter and emergency medical practitioner. Engaging in a career transition, John began working for the National Park Service while pursuing an MS in Recreation Administration. John's experience working in Antarctica fostered an interest in visitor codes of conduct especially as they pertain to the polar regions. He currently sits as co-chair for the US Association of Polar Early Career Scientists and is a national committee representative for the Association of Polar Early Career Scientists.

Milena Manhães Rodrigues, PhD student in Tourism, University of São Paulo, São Paulo, Brazil. Milena is researching community-based tourism with traditional communities. She received a masters degree in Community Psychosociology and Social Ecology at the Federal University of Rio de Janeiro (EICOS/UFRJ). She is currently a Pedagogical Distance Mediator in the discipline of Tourism and Society at Rural Federal University of Rio de Janeiro (UFRRJ). She also works as a researcher for the project 'Sanitary, Socio-environmental and Health Conditions', a cooperation between Funasa and Fiocruz.

Chris Ryan, Foundation Professor, University of Waikato, Hamilton, New Zealand. Chris is the Director of the BUU China–New Zealand Tourism Research Unit. He spearheaded a UNWTO research base studying Raglan under the UNWTO INSTO programme. An emeritus editor of *Tourism Management*, he also founded *Tourism Management Perspectives* and is currently editor of *Tourism Critiques: Practice and Theory*. Chris was noted by PLOS ONE as being one of the top 1% of global researchers.

Renato de Oliveira dos Santos, PhD student in Social Change and Political Participation at the University of São Paulo (USP). Substitute Teacher at the Institute of Applied Social Sciences of the Federal Rural University of Rio de Janeiro – UFRRJ. Master in Community Psychosociology and Social Ecology at the Institute of Psychology of the Federal University of Rio de Janeiro - Bachelor in Tourism at the Federal Rural University of Rio de Janeiro and Bachelor in Administration. She develops research and actions directed to the following related themes: Sustainability, Community Based Tourism, Traditional Peoples and Communities, Local Development, Collective Processes, Protected Areas, Memory and Narrative. Researcher: Tourism Teaching and Research Center; Memory, Territories and Occupations Laboratory: Sensitive Tracks and Laboratory of Community-Based Tourism, Sustainability and Networks (CBT-Networks).

Jessica A. Schottanes, Jess is a Landscape Designer with experience developing open space and recreation plans, researching mountain and trailside communities, exploring early 20th-century sustainability sensibilities, and designing gateway destinations. After completing her master's thesis on Emile Benton MacKaye's Original Vision of Regional Planning, she was inspired to move to Appalachia. Today, Jess holds a Bachelor of Science in Environmental Geoscience from Slippery Rock University as well as a Master of Science in Geography and Master of Regional Planning from the University of Massachusetts, Amherst. When Jess isn't in the studio, you can find her reading about trailblazing women and lost ski areas or brushing up on her cartographic and natural science illustration skills.

Ole R. Sleipness, Associate Professor, Department of Landscape Architecture and Environmental Planning, Utah State University, Logan, Utah, USA. Ole received his PhD in Environmental Design and Planning from Clemson University in 2009. His research interests include rural community development, the interface between public and private lands and the ways university design programmes engage these dynamic landscapes, particularly in the Intermountain West. In his recreation design and open space studio, he enjoys integrating his research expertise while providing students with opportunities to address real-world problems through community engagement projects.

Susan L. Slocum, Associate Professor, Tourism and Events Management, George Mason University, Virginia, USA. Sue has worked on regional planning and development for 15 years and with rural communities in Tanzania, the United Kingdom, Belarus and the USA. Her primary focus is on rural sustainable development, policy implementation and food tourism, specifically working with small businesses and communities in less advantaged areas. Sue received her doctoral education from Clemson University and has worked at the University of Bedfordshire, UK and at Utah State University. She was awarded the Fulbright Scholarship in 2020.

Maria A. Ventura, Assistant Professor, Department of Biology, University of the Azores, Ponta Delgada, Portugal. Maria has a Biology degree from the University of Lisbon and gained a PhD in Animal Ecology from the University of the Azores in 2003. She teaches classes on conservation biology, sustainable tourism in protected areas and conservation of species and habitats, to undergraduate, masters and PhD students in Biology. Her main research interests are conservation, impacts of tourism on biodiversity and valuing protected areas for nature-based tourism. She has supervised several graduate, masters and PhD students, and has also been PI or participant in several research projects.

Therez B. Walker, lecturer, Centre for Hotel and Tourism Management, University of the West Indies, Mona, Jamaica. Therez's research work focuses on sustainable tourism development in Small Island Developing States (SIDS), and how islands can maximize linkages between tourism and other sectors of the local economy. In addition to academic activities, she has a background in consultancy, working with destination management organizations (DMOs) and non-governmental organizations (NGOs).

Peter Wiltshier, Professor Emeritus, University of Derby, Derby, UK. Peter explores responsible development of tourism in a creative and dedicated fashion, supporting communities to provide resources for their welfare and dedicated to the pursuit of a responsible future with a focus on the beautiful and important 'Green Lungs' of rural Britain in the East Midlands, as well as in New Zealand. His role as Senior Lecturer, Destination and Community Tourism Management at the University of Derby is to ensure that the public and private sector work together to develop resources and skills for communities to take charge of their own destinies. Peter also works with the Diocese of Derby to identify how tourism can benefit churches. He endeavours to create a better environment for all through purposeful leisure and recreation.

1 Introduction

Peter Wiltshier[1]* and John B. Read IV[2]
[1]University of Derby, Derby, UK; [2]University of Waterloo, Ontario, Canada

Abstract
Our introduction delivers an overview of the challenges facing gateway communities surrounding protected areas (PAs) and provides brief descriptions of the chapters within this book. It is intended to situate the chapters in relation to previous literature and showcase the importance of new and innovative approaches to community engagement for those most impacted by the protective status of valuable natural resources.

Keywords: gateway communities, protected areas (PAs), conservation, community engagement

1.1 Book Overview

It is our pleasure to present 11 highly detailed case studies from gateways of protected area (PA) sites across the globe. The commonality across these studies is the extent to which stakeholder perspectives are being used, as well as the competing and conflicting demands of public and private stakeholders dealing with the contradictory needs of consumers and the communities hosting visitors. Mandić and Petrić (2021, p. 1) explain that 'protected areas ... are clearly defined geographical spaces that are recognised as such and are dedicated to achieving long-term nature conservation, with the associated ecosystem services and cultural values'. This book focuses primarily on gateway communities – those that share a border with a PA, provide entrance and exit points to the attractions and are inextricably linked through common geography, conservation resources and tourism visitation impacts (Burghardt, 1971). These relationships are often contentious, and our goal is to provide mutually beneficial alternatives to PA–community relationships that support both the sustainability of the PAs themselves, as well as community wellbeing.

Policy and philosophy guiding national park management globally has been significantly influenced by North American perceptions and ideals on what constitutes 'wilderness'. This means that national park management in the USA was established when engagement with local communities was considered and/or deemed as less of a priority than resource protection, which made effective collaborative decision-making a one-sided affair and resulted in limited cooperation between stakeholders. Today we must reflect on this history while also exploring the instances of PA management that predate those established in the USA (e.g., Main Ridge Forest Reserve in Tobago, 1776 and Bodg Khan Uul in Mongolia, 1778, among others). The goal is to develop strategies and policies from central and local governance perspectives that provide both philosophical and practical guidelines for management and an inclusive focus for tourism and related micro-business, to

*Corresponding author: pwiltshier@yahoo.co.nz

DOI: 10.1079/9781789249033.0001

support gateway communities that can thrive independently from conservation and the physical constraints on PA development.

There are diametrically opposed perspectives on best practice for managing and developing PAs. We consider the two approaches to identifying and promoting PA gateways. One engages governance and management; the second identifies a socially just and environmental approach without an explicit focus on governance and management. Our case studies take both approaches. The governance and management of protected areas relies heavily on the use and interpretation of the judicial and legislated needs of environments. In the second approach, observation of what works for a community or destination and its stakeholders is without doubt pragmatic.

However, evidence from case studies and empirical research clarifies that there are multiple approaches that governments, scientists, students, stakeholders and local community members may take to identifying best practices in sustainable development for communities at the gateways to PAs. There are many conceptual approaches to the study of PAs and their important gateways. These include a critical political ecology approach (Jamal and Stronza, 2009), a complex adaptive system focus (Schianetz and Kavanagh, 2008), participatory approaches in lieu of expert-based approaches (Langton *et al.*, 2014) and political ecology, as discussed by Bohn and Hall in Chapter 3, among others. The case studies in this book have produced evidence that interventions can be used to accrue cohesive, enduring and sharing practices that underpin the contemporary United Nations (UN) established Sustainable Development Goals (SDGs) (2015) for the overarching deployment of achieved knowledge.

PAs have often been created around the core value of 'wilderness', which has historically involved the displacement of Indigenous peoples, and others, living on the land that is to be protected (Chatty and Colchester, 2002). Modification of landscapes and construction efforts in these areas are often limited to approved projects by federal government agencies tasked with managing such areas (e.g., National Park Service in the USA). However, alternative approaches that have seen mixed success in the USA involve partnerships with varying levels of independence in regard to maintaining PA landscapes (discussed in Chapters 2 and 10). At one extreme, Bristow, Judkins and Schottanes (Chapter 2) discuss the original role of community members within multiple gateway communities along the 3200-kilometre (2000-mile) Appalachian Trail, as stewards whose engagement extends beyond just their local gateways that line the trail, to the creation and maintenance of the trail since its inception. While Sleipness, Powell and Curtis (Chapter 10) discuss how public–private partnerships can play a beneficial role in bringing together communities and the agencies that manage PAs to address the needs of both, especially as visitation rates continue to increase to PAs, specifically national parks. In both of these cases, and as demonstrated in other chapters of this volume, non-profit and other non-governmental organizations (NGOs) can play a significant role in mediating the relationship between government agencies and local communities to co-create mutual benefits.

NGOs, especially those with an environmental focus, can also play an alternative role of natural area and species protection, as seen in Chapters 5 and 7. The diametrically opposed visions of environmentalists and socioeconomic groups is emphasized by the contestation of spaces and places for development, that may well vary over time and be highly dependent on externalities, such as trade, politics and global climate change concerns. However, NGOs, conservation organizations and other non-profit groups can advocate for environmental protection, but are often reliant upon non-guaranteed grant funding (Sullivan, 2002) that disenfranchises gateway communities. Especially in those areas that are composed of primarily Indigenous peoples, contemporary management practices may result in the degradation of an environment that has seen centuries of traditional land management practices (Chatty and Colchester, 2002). However, social and economic development should not always be viewed through a Eurocentric perspective. As decolonial tourism literature highlights (e.g., Tebrakunna Country and Lee, 2017), the direction of social and economic development must be driven by Indigenous communities.

The two primary and opposed purposes for PA gateways are conservation and development:

conservation for species protection; and development for sustainable protection and enhancement of social and economic features. This conceptual dual focus is typical of the co-management approach (Berkes, 2009; Gruber, 2010) and is related by Walker and Lawrence in Chapter 5. Yet, within these two diametrically opposed purposes and agendas, there are exclusionary conservation and biodiversity claims (Lele *et al.*, 2010). To minimize disparate approaches and conflicting issues, the opportunity exists to employ conceptual systems thinking or process-orientated resilience systems thinking (Strickland-Munro *et al.*, 2010). Fig. 1.1 reminds the reader that interventions may be borrowed from other disciplines to test and identify solutions that guarantee protection and ensure viability within the context of sustainable development.

In turn, this drives an interest in reducing conflict between stakeholders by adopting the community-driven approach (Waterton and Watson, 2013). Competing and comparative frames of examination for PA visitation can blend stakeholders' varying perspectives and goals in setting the agenda for a community approach. Community-driven visions are espoused in Chapters 3 and 9. Taking a step further, it is possible to grow capacity by building new opportunities to participate in skill development training and related new business opportunities, allowing participation in income-generating activities. Scoping and costing include identification of the levels of involvement by key stakeholders, a description of the context and identification of key conservation goals. Data is collected using socioeconomic variables and threats to biodiversity and natural features. Conservation objectives are set, achievements are reviewed, then additional conservation areas are selected. Conservation actions are then applied to maintenance and monitoring procedures (Andrade and Rhodes, 2012).

At a scale that is appropriate for protected, and largely unpopulated, areas we observe the need for, and availability of, small loans to help key stakeholders in the development of sustainable livelihoods and asset-based approaches (Bennett *et al.*, 2012). In Chapter 9, the capital assets framework, which may be required for protecting a variety of community assets into the future, is discussed. The framework includes: defined roles and responsibilities; cross-interest stakeholder participation; public participation to determine defined goals, objectives, and decision/support-making systems (Gleason *et al.*, 2010); and cross-disciplinary approaches (McCool, 2009; Redpath *et al.*, 2013). It

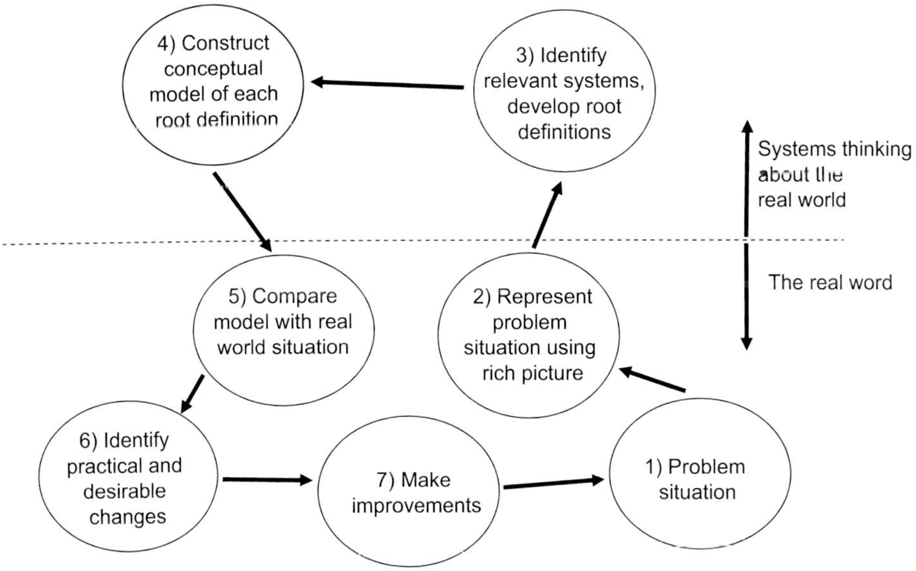

Fig. 1.1. Conceptual or process-orientated resilience systems thinking (adapted from Checkland, 1999).

highlights messy and uncertain processes and outcomes relating to equitable access to knowledge obtained through partnerships. PA tourism planning partnerships are characterized by certain attributes: representativeness; a sense of ownership; a learning focus; and attention to relationships. However, these attributes are not easily attained, for they are facilitated by several contextual factors and processes, such as trust, political and economic power and equitable access to knowledge (McCool, 2009).

Community and multi-level approaches that empower and engage a variety of outcomes and actions (Lockwood, 2010) are addressed by Ventura, Costa and Botelho in Chapter 7. There is a demonstrable focus on integrated landscape approaches (Reed *et al.*, 2016). Petrić and Mandić (2021, p. 15) write, 'one of the solutions might be a transition from the traditional top-down governance of protected area (PA) to adaptive co-management (ACM), which advocates participation, collaboration, and iterative learning'. Another avenue of consideration is a stakeholder approach, as advocated by Pomeroy and Douvere (2008). Again, we consider diametrically opposed perspectives of protection and gateways depending on the view taken by the stakeholder. Government adherents will necessarily take the perspective of their political campaigns, that for which they were elected by voters in a democracy. Stakeholders with a profit motive will be less inclined to altruism and philanthropy and more focused on neoliberal democratic perspectives. That is not to say that there isn't room for both standpoints – it is more important to consider the long-term sustainability of the stakeholder's view and focus on, perhaps, the SDGs (United Nations, 2015).

In the modern, Anthropocene era, consumers and suppliers can take a shared and cooperative view of the task of protecting spaces through better information dissemination. That is one reason this text is relevant and enlightening. Citizen responsibility is in vogue; it is the charge we all make upon our fellow connected stakeholders, and it represents itself in many forms (Conrad and Hilchey, 2011). Volunteers in citizen science adopt a 'can do' profile and undertake empirical work to improve our knowledge and then spread this knowledge through voluntary data collection within a community, as well as through the academy. This citizen-based approach is related in Chapter 10 via Zion National Park citizen science approaches that support the social and ecological resilience represented by community-based tourism, which is 'a new approach to the study of socio-ecological sustainability, affirming that sustainable development can only be achieved in sufficiently resilient socio-ecosystems' (Ruiz-Ballesteros, 2011, p. 665). Cooperative and citizen-based stakeholders need empowerment, access to information, access to conservation education programmes and resources to help plan and support decision-making processes.

Equity in decision-making is emerging in less developed locations as critical in the sustainability of management (Chapters 8 and 9 are representative). Bottom-up or endogenous concerns and issues are perceived as critical in outcomes that must be continuously supported when compared to central or top-down dictats. The studies contained here imply that new threads of issues connecting stakeholders in feedback loops and in decision support processes determine external or internal participation (Dawson *et al.*, 2018). Environmental justice and participatory processes to ensure best practices, which emerge from global telecommunications and knowledge sharing, are critical and a legacy of the studies contained here. 'Telecoupling', as a phenomenon, can play a bigger part in the accrual and sharing of this new knowledge in order to maintain sustainability in different protected areas (Boillat *et al.*, 2018). Typically, the International Union for Conservation of Nature (IUCN) offers stakeholders a framework for adopting protocols and managing protected area destinations with conceptual equity in gateway and hinterland sustainable development PA guidelines (Vasilijević *et al.*, 2015).

What appears to be important for gateway communities, in the context of supporting PAs, is the economic benefits represented in employment opportunities in tourism businesses (Chapter 3), opportunities for micro-enterprises, and collective economic benefits. Citizen science approaches are delivered to support biodiversity conservation and related environmental services and participation in biodiversity conservation programmes (Chapters 2 and 3). Amenities development and local infrastructure for site protection include conservation infrastructure and related demand-driven tourism superstructures and emergent resilient tourism products (Nyaupane and Poudel, 2011).

Christaller's (1972) view of competition for scarce resources and locating, spatially, the tourism opportunities in a socioeconomic and environmentally sustainable manner has a pedigree with relevance to the debate over gateway communities for PAs. Gateways occupy site-specific and purposeful roles that provide services to surrounding PAs. Christaller's 'central place theory' has given rise to a slew of conceptual approaches that deliver sustainable environmental outcomes in PAs. Such approaches include VIM (visitor impact model), LAC (limits of acceptable change), VAMP (visitor activity management process) and TOMM (tourism optimization management model) (Graefe *et al.*, 1985). The key to these concepts is the ongoing monitoring and managing of PAs through a stakeholder group, universally admired by governance and by socioeconomic groupings, with responsibilities identified within the model for meeting targets, such as the vaunted SDGs model espoused through the United Nations' vision of 2015. Dela Santa, Ramos-Tumanan and Ryan use the Batanes, Philippines, to update our perception of shared and partnership public–private cooperation (Chapter 8).

Species protection accompanies natural landscape conservation and management as can be seen in Chapter 5 on Antigua and Barbuda. Stakeholders with determination can put forward a strong case for endangered species to be the centre of management agendas and development. Protecting natural landscapes with an identifiable mission developed from the core, or gateway, can help formulate the important role that the periphery owes to committed stakeholders at the core. It is essential that legislation and management from the core exists to support the aim to place species conservation at the heart of any marketing and promotional activity for the community.

It is not only natural area conservation that is important. The preservation of unique sociocultural heritage is of equal concern in gateways of intense consumer demand, such as Magnetic Island in Queensland, Australia. In Chapter 11, Moscardo draws our attention to dual roles: hosts of environments and hosts of heritage preservation and interpretation (while Chapter 8 highlights the constraints of natural area conservation on the preservation of traditional livelihoods and heritage of the Batanes). Stakeholders' endangered landscapes, those with specific limitations on consumer exploitation and extensive marketing, acknowledge communities bounded by infrastructure that can barely cope with host community demands. Pollution from rubbish and sewage can demonstrably limit exploitation by stakeholders in situations of precariously balanced supply and demand, such as that in Paraty, Brazil, as indicated by Rodrigues, dos Santos, Raimundo and Machado in Chapter 9. By contrast, urban environments struggle to expedite PA legislation under the most extreme consumer demand pressures. In Chapter 12, Daniels and Slocum show that The National Mall, Washington DC struggles to maintain legitimacy within the limits of national park legislation and the expectations of rampant consumption by visitors. The concept of the PA gateway takes on a different and complex mantle with the paradoxical demands of visitors and the heritage landscape described in Chapter 12. Moreover, in Western North America, the opportunity exists to evaluate new gateway communities protecting existing and mature national parks. The example of Zion, in Chapter 10, provides evidence of citizen science and students engaged in research to identify good practices, showing that an emphasis on providing alternative gateways predicated on differing demands from consumers is now available.

Inclusivity and the next generation are covered in Chapters 2 and 3. Chapter 2, on the Appalachian Mountains in Eastern North America, discusses how a planned multi-state hiking trail was initially designed around gateway communities acting as partners, including involvement in trail creation and maintenance. It then discusses how current perceptions of community members in many of these gateway cities, and the efforts of state and non-governmental actors, may shape the future of community engagement with the Appalachian Trail in the future. Conversely, by engaging with a political economy approach, Chapter 3 discusses the changes that have occurred within Finland, and specifically the Lapland region, which has encouraged an Arctic identity as a driver of tourism. This demonstrates a shift from an area that operated as small communities, where Indigenous Sami maintained traditional livelihoods, to one that the authors identify as a 'spatial re-imagination of Lapland in the Arctic [to] represent another neoliberal step towards the commodification of the environment' – driven by state and corporate actors.

Evidence suggests that there are evolving relationships that join conservation and community engagement over time. In Chapter 4, Lackey, Bricker and Hicks discuss the evolution of partnerships and communication between a federally operated national park (Grand Tetons), the local community of Jackson, and park concessionaires. By highlighting insights from individuals with leadership roles within each entity, the authors demonstrate that joint efforts have increased over time as the national park has shifted from operating as an 'island', disconnected from surrounding communities, to being a more engaging partner. However, the chapter shows that there is still room for growth as government entities can require large commercial investment to drive movement towards greater stakeholder involvement.

Strengthening the local community involves emphasis on developing new capacities that are required as a result of tourism, both to manage conservation zones and to undertake remedial work under the SDGs, which in turn requires investment in resources for learning and education. In Chapter 6, we see the strengths that come from gateway community involvement in the development of sustainable tourism to a destination (Samalayuca Dunes, Mexico). Through the involvement of individual community members, González-Herrera, Suárez-Chaparro and Hernández-Casimiro demonstrate how research can assist in achieving both the SDGs and the 'three pillars of ecotourism' (natural/cultural heritage; economic development; and social wellbeing). The authors then generate both a strategy and action items for achieving sustainability through gateway community involvement that can serve as a model for PAs engaging with communities.

Cross-border and shared partnership approaches to tourism and conservation are highlighted by Ventura, Costa and Botelho in Chapter 7 through the assessment of three small Atlantic archipelagos – Cabo Verde, Fernando de Noronha and Azores. The authors discuss how environmental non-governmental organizations (ENGOs) have the potential to serve as intermediaries between the gateway communities within which they exist and governmental organizations, especially in regard to natural area protection on islands. This chapter further explores the difficulties that ENGOs can experience when operating on islands and how this can limit their ability to both involve local communities as well as advocate for sustainable practices.

> There is a pertinent need for the development of knowledge that enables us to close the considerable gap that continues to exist between managerial responsibility and practice on the organizational frontline. Across many industries ... a relentless series of scandals shows us that the pursuit of self-serving behaviour continues to take precedence over collective interests and well-being. This emphasizes the need for managerial practice to adopt a stewardship perspective.
>
> (de Ruyter and Keeling, 2019, np)

There is ample evidence that 'more of the same' thinking in conservation and protection has not yet produced results to overcome the volatility of our existence, nor does it deal with the complexities that better access, more inclusivity and acknowledged diversity can offer all generations. As Moscardo highlights in Chapter 11, 'storying' through co-creation with community members can allow for the development of stories that incorporate both tradition and ideals for future tourism within gateway communities. This can result in telling the narratives that communities wish to communicate, not those generated by outsiders or marketing entities.

1.2 Concluding Remarks

In light of globalization, improved communications and technology, and demand for equity, inclusivity and accessibility, the accelerating pace of change has indeed become obvious to all stakeholders. The neoliberal market forces model inherent in Keynesian economics and reinforced by western leaders, such as Reagan and Thatcher in the 1980s, has increased incrementally the speed of change and pressure to speed up everything in science, business, technology, medicine, government, finance and development, as well as in individual personal routines (Job *et al.*, 2017). We need to challenge conventional thinking in this post-industrial era. Connectivity is possible across many topics and issues, and encourages us to understand change and its impact today (Westphal in Schroeter, 2020, p. 11). Global competition, increased depredation of the physical environment for economic gain, a failure to grasp alternative ways of being against a background typified by improved science, technology and communications, have all caused us all to reconsider and reflect on governance and development. Unfortunately, this reflection has taken more than a generation, and there is now angst and guilt expressed by the baby-boomers as well as generations X, Y and Z. We are confronted by uncertainty and anxiety about *quo vadis*. This text demands that we reflect on convention and challenge traditional ways of governance and development.

We acknowledge the need for better, but maybe not greater, productivity. We express sensitivity to the needs of the dispossessed, and we hope a voice for social justice is now heard. An algorithm to support social justice, to minimize the market-force economics, is emergent. To address irrational commercial demands, to truly comprehend extinction of species and the alteration to landscapes beyond repair, we need greater adaptability; we need resilience in products and services, to be consumed equitably. As noted by Read and Grimwood (2021), when engaging with communities for tourism development, there is also a need to continue incorporating emerging ethical thought to counter traditional forms and methods of development. New knowledge, coupled to technology, and the science of better productivity can connect global citizens to choices where non-linear growth and actual collectivism can be the focus. As Watson (cited in Schroeter, 2020, p. 138) states, 'there is a need for us to augment the ability and capacity to make good moral judgments based on the science of new algorithms embedding fifty years of learning'.

References

Andrade, G.S.M. and Rhodes, J.R. (2012) Protected areas and local communities: an inevitable partnership toward successful conservation strategies? *Ecology and Society* 17(4), 1–16. DOI: 10.5751/ES-05216-170414.

Bennett, N., Lemelin, R.H., Koster, R. and Budke, I. (2012) A capital assets framework for appraising and building capacity for tourism development in aboriginal protected area gateway communities. *Tourism Management* 33(4), 752–766. DOI: 10.1016/j.tourman.2011.08.009.

Berkes, F. (2009) Evolution of co-management: role of knowledge generation, bridging organizations and social learning. *Journal of Environmental Management* 90(5), 1692–1702. DOI: 10.1016/j.jenvman.2008.12.001.

Boillat, S., Gerber, J.-D., Oberlack, C., Zachringer, J.G., Ifejika Speranza, C. *et al.* (2018) Distant interactions, power, and environmental justice in protected area governance: a telecoupling perspective. *Sustainability* 10(11), 3954. DOI: 10.3390/su10113954.

Burghardt, A.F. (1971) A hypothesis about gateway cities. *Annals of the Association of American Geographers* 61(2), 269–285. DOI: 10.1111/j.1467-8306.1971.tb00782.x.

Chatty, D. and Colchester, M. (eds) (2002) *Conservation and Mobile Indigenous Peoples: Displacement, Forced Settlement, and Sustainable Development*. Berghahn Books, New York, NY, USA.

Checkland, P. (1999) *Systems Thinking, Systems Practice*. John Wiley & Sons Ltd, Chichester, UK.

Christaller, W. (1972) How I discovered the theory of central places: a report about the origin of central places, man, space, and environment. In: *Concepts in Contemporary Human Geography, Red*. PW English, RC Mayfield, New York–London–Toronto, pp. 601–611.

Conrad, C.C. and Hilchey, K.G. (2011) A review of citizen science and community-based environmental monitoring: issues and opportunities. *Environmental Monitoring and Assessment* 176(1–4), 273–291. DOI: 10.1007/s10661-010-1582-5.

Dawson, N., Martin, A. and Danielsen, F. (2018) Assessing equity in protected area governance: approaches to promote just and effective conservation. *Conservation Letters* 11(2), e12388. DOI: 10.1111/conl.12388.

de Ruyter, K. and Keeling, D. (2019) Responsible management on the organizational frontline: a stewardship perspective. *Emerald Open Research* 1, 6. DOI: 10.12688/emeraldopenres.12908.1.

Gleason, M., McCreary, S., Miller-Henson, M., Ugoretz, J., Fox, E. *et al.* (2010) Science-based and stakeholder-driven marine protected area network planning: a successful case study from north central California. *Ocean & Coastal Management* 53(2), 52–68. DOI: 10.1016/j.ocecoaman.2009.12.001.

Graefe, A.R., Kuss, F.R. and Loomis, L. (1985) *Visitor impact management in wildland settings.* Proceedings-National Wilderness Research Conference-current research, Fort Collins, CO, July 23–26, 432–439.

Gruber, J.S. (2010) Key principles of community-based natural resource management: a synthesis and interpretation of identified effective approaches for managing the commons. *Environmental Management* 45(1), 52–66. DOI: 10.1007/s00267-008-9235-y.

Jamal, T. and Stronza, A. (2009) Collaboration theory and tourism practice in protected areas: stakeholders, structuring and sustainability. *Journal of Sustainable Tourism* 17(2), 169–189. DOI: 10.1080/09669580802495741.

Job, H., Becken, S. and Lane, B. (2017) Protected Areas in a neoliberal world and the role of tourism in supporting conservation and sustainable development: an assessment of strategic planning, zoning, impact monitoring, and tourism management at natural World Heritage Sites. *Journal of Sustainable Tourism* 25(12), 1697–1718. DOI: 10.1080/09669582.2017.1377432.

Langton, M., Palmer, L. and Rhea, Z.M. (2014) Community-oriented protected areas for indigenous peoples and local communities. In: Stevens, S. (ed.) *Indigenous Peoples, National Parks, and Protected Areas: A New Paradigm Linking Conservation, Culture, and Rights*. The University of Arizona Press, Tucson, AZ, USA, pp. 84–107.

Lele, S., Wilshusen, P., Brockington, D., Seidler, R. and Bawa, K. (2010) Beyond exclusion: alternative approaches to biodiversity conservation in the developing tropics. *Current Opinion in Environmental Sustainability* 2(1-2), 94–100. DOI: 10.1016/j.cosust.2010.03.006.

Lockwood, M. (2010) Good governance for terrestrial protected areas: a framework, principles and performance outcomes. *Journal of Environmental Management* 91(3), 754–766. DOI: 10.1016/j.jenvman.2009.10.005.

Mandić, A. and Petrić, L. (2021) Introduction to Mediterranean protected areas in the era of overtourism. In: Mandić, A. and Petrić, L. (eds) *Mediterranean Protected Areas in the Era of Overtourism*. Springer, Cham, Switzerland, pp. 1–12.

McCool, S.F. (2009) Constructing partnerships for protected area tourism planning in an era of change and messiness. *Journal of Sustainable Tourism* 17(2), 133–148. DOI: 10.1080/09669580802495733.

Nyaupane, G.P. and Poudel, S. (2011) Linkages among biodiversity, livelihood, and tourism. *Annals of Tourism Research* 38(4), 1344–1366. DOI: 10.1016/j.annals.2011.03.006.

Petrić, L. and Mandić, A. (2021) Governance and management of protected natural areas in the era of overtourism. In: Mandić, A. and Petrić, L. (eds) *Mediterranean Protected Areas in the Era of Overtourism*. Springer, Cham, Switzerland, pp. 15–24.

Pomeroy, R. and Douvere, F. (2008) The engagement of stakeholders in the marine spatial planning process. *Marine Policy* 32(5), 816–822. DOI: 10.1016/j.marpol.2008.03.017.

Read, J. and Grimwood, B.S.R. (2021) Ecotourism development through culturally sensitive universalism. In: Fennell, D.A. (ed.) *Routledge Handbook of Ecotourism*, 1st edn. Routledge, London, UK, pp. 103–116.

Redpath, S.M., Young, J., Evely, A., Adams, W.M., Sutherland, W.J. *et al.* (2013) Understanding and managing conservation conflicts. *Trends in Ecology & Evolution* 28(2), 100–109. DOI: 10.1016/j.tree.2012.08.021.

Reed, J., Van Vianen, J., Deakin, E.L., Barlow, J. and Sunderland, T. (2016) Integrated landscape approaches to managing social and environmental issues in the tropics: learning from the past to guide the future. *Global Change Biology* 22(7), 2540–2554. DOI: 10.1111/gcb.13284.

Ruiz-Ballesteros, E. (2011) Social-ecological resilience and community-based tourism: an approach from Agua Blanca, Ecuador. *Tourism Management* 32(3), 655–666.

Schianetz, K. and Kavanagh, L. (2008) Sustainability indicators for tourism destinations: a complex adaptive systems approach using systemic indicator systems. *Journal of Sustainable Tourism* 16(6), 601–628. DOI: 10.1080/09669580802159651.

Schroeter, J. (2020) *After Shock: The World's Foremost Futurists Reflect on 50 Years of Future Shock.* Abundant World Institute, Bainbridge, WA, USA.

Strickland-Munro, J.K., Allison, H.E. and Moore, S.A. (2010) Using resilience concepts to investigate the impacts of protected area tourism on communities. *Annals of Tourism Research* 37(2), 499–519. DOI: 10.1016/j.annals.2009.11.001.

Sullivan, S. (2002) How sustainable is the communalizing discourse of "new" conservation? The masking of difference, inequality and aspiration in the fledgling "conservancies" of Namibia. In: Chatty, D. and Colchester, M. (eds) *Conservation and Mobile Indigenous Peoples: Displacement, Forced Settlement, and Sustainable Development*. Berghahn Books, New York, NY, USA, pp. 158–187.

Tebrakunna Country and Lee, E. (2017) Performing colonisation: the manufacture of Black female bodies in tourism research. *Annals of Tourism Research* 66(37/38), 95–104. DOI: 10.1016/j.annals.2017.06.001.

United Nations (2015) *Transforming our world: the 2030 agenda for sustainable development*. Available at: https://www.un.org/ga/search/view_doc.asp?symbol=A/RES/70/1&Lang=E (accessed 1 February 2022).

Vasilijević, M., Zunckel, K., McKinney, M., Erg, B. and Schoon, M. (2015) *Transboundary Conservation: A Systematic and Integrated Approach*, Vol. Best Practice Protected Area Guidelines Series No. 23. IUCN, Gland, Switzerland, pp. xii–107.

Waterton, E. and Watson, S. (2013) Framing theory: towards a critical imagination in heritage studies. *International Journal of Heritage Studies* 19(6), 546–561. DOI: 10.1080/13527258.2013.779295.

2 Appalachian Trail Communities: Gateway Partners in Parks and Protected Area Management

Robert S. Bristow[1]*, Julie Judkins[2] and Jessica Schottanes[3]
[1]*Westfield State University, Massachusetts, USA;* [2]*Appalachian Trail Conservancy, North Carolina, USA;* [3]*University of Massachusetts, Massachusetts, USA*

Abstract
The Appalachian National Scenic Trail is a 2200-mile footpath in the mountains of the Eastern United States. In 1921, regional planner Benton MacKaye envisioned a trail that would support sustainable land management strategies that required an investment of local volunteers. By 1937, a continuous route was blazed from Maine to Georgia. Engaging trailside communities in active management, sustainable promotion and decision-making is crucial to protect the Trail and its diversity of resources. Two online surveys were distributed to the trailside communities in a partnership programme to identify how citizens embraced the responsibility of being gateway partners and the mutual benefits received. This research examines what we can learn from MacKaye's model to build on the local linkages for parks and protected areas (PAs) and compares it to contemporary challenges. Given the Trail's length, multiple ecosystems and economies exist along the route, unlike singular gateways around more traditional parks and PAs.

Keywords: Appalachian Trail, gateway communities, parks and protected areas (PAs), national parks, trail towns

2.1 Introduction

The Appalachian National Scenic Trail (Trail) is known worldwide for its 2200 miles of walking footpaths on the ridgelines of the Appalachian Mountains in the Eastern United States. However, when Benton MacKaye, who first wrote and envisioned the footpath in his article 'The Appalachian Trail: A project in regional planning' for the *Journal of the American Institute of Architects* in 1921, his vision encompassed much more than just a long-distance hike. He proposed a sphere that could be a sustainable and regenerative land management strategy as an impetus for social change, requiring a culture of local investment to promote recreation, recuperation, scientific investigation and outdoor education. Shortly after his writing, MacKaye and other like-minded environmentalists established the Appalachian Trail Conference (henceforth ATC and now Conservancy) in 1925, which organized east coast volunteer hiking clubs to build the new Trail. By 14 August 1937, a continuous route existed from Maine to Georgia.

In its early days, the Trail remained vulnerable and unprotected by federal and state regulations. From the 1930s, the Trail route had

*Corresponding author: rbristow@westfield.ma.edu

 Tourism Transformations in Protected Area Gateway Communities (eds S.L. Slocum *et al.*)
DOI: 10.1079/9781789249033.0002

to be frequently changed due to urban sprawl and competing land-use demands (Fisher and Durrance, 1972; Foster, 1987). By the 1960s, frequent relocations and landowner disputes riddled the route of the Trail. The historic 'handshake' agreements between the Trail volunteers and landowners meant the Trail's future was threatened (Bureau of Outdoor Recreation, 1966). Yet, at the same time, the Trail was gaining popularity with upwards of 1 million visitors per annum.

In 1968, after a sweeping lobbying effort by the Conservancy, President Johnson signed the National Scenic Trails Act into law, putting the Appalachian National Scenic Trail under federal protection as a National Park Unit and formulating a long-standing collaboration with local Trail groups as management partners (King, 2000). Currently, 31 volunteer trail clubs have remained vital stakeholders in the Trail's development, advocacy, management and protection.

MacKaye's work envisioned local towns as necessary partners in the Trail's legacy and experience, acting as key contributors and managers. Trailside communities' participation in active management, promotion and decision-making is crucial to protect the Appalachian National Scenic Trail and the diversity of resources surrounding the Trail, as community participation in decision-making within protected areas (PAs) is the most significant factor in community compliance with public land policies (Howe *et al.*, 1997; Judkins, 2015; Bello *et al.*, 2017; Schottanes, 2021).

This chapter explores the current status and challenges that exist between the Trail, the gateway communities and the volunteers and partners obligated to manage it, as well as the interconnections between the three. A literature review will cover an introduction to parks and PA management. Gateway communities are described next, along with their unique and reciprocal contribution to parks and PAs. Next, a history of the Appalachian Trail will trace the initial long-distance Trail's vision through to today's multi-faceted management concerns. Findings will be shared in the results section, followed by a discussion. The chapter will conclude with future needs and additional considerations in gateway communities, parks and PAs.

2.2 Literature Review

2.2.1 Parks and PA management

Sustainably and adequately managing parks and PAs is a challenge crucial to the long-term survival of natural and cultural treasures (Eagles and McCool, 2002; Stevens, 2014; Leung *et al.*, 2018). The variety of landscapes, cultural differences and management styles complicate the mission of land managers who are responsible for protecting all the resources within the area. Establishing PAs and parks is a relatively recent development, and governments continue to face the need to manage these lands and waters. The largest national park in Europe, Vatnajökull National Park, was newly created in 2008; even in the United States, with its abundant quantity of parklands, the history of protection is only just over 100 years old. This lack of historical precedent leads to a need to play 'catch up' on management strategies for parks and PAs.

MacKaye described his philosophy in *The new exploration* as the need for common public grounds to remind us of the goal of living: so that work and art and recreation and living will all be one (MacKaye, [1928] 1990). Selin *et al.* (2020) acknowledge the importance of public lands and the multiple benefits to local communities. New residents are moving to rural areas based on environmental and scenic values (Jones *et al.*, 2003). The challenge is to bridge the gap between economic development from outdoor recreation and natural park assets within local communities and the appropriate management level for the parks and PAs. This dual nature, between local towns and parks, is essential. It includes job creation, consumer expenditure and new business growth, all supported by the presence of a local public park that provides a visitor destination (Cerveny *et al.*, 2020).

Numerous tasks face communities and park managers when creating sustainable plans and shared visions for park resources and their future desired conditions. In 1994, the US Government Accountability Office (US Government Accountability Office (GAO), 1994) found that the US Park Service had no complete inventory of external threats, sources of the threats or mitigation of the threats. This lack of data meant, for example, that there was no practical means of working with outside parties. The GAO reported four main threats to the nation's national parks: urban encroachment; water quantity and quality issues; air pollution; and human activities. While partially addressed in subsequent years, all parks and PAs continue to be threatened by the same concerns today.

Many of these pressures impact sustainable tourism in gateway communities. One long-standing strategy is building partnerships between park areas and outside interests (President's Commission on Americans Outdoors, 1987). For example, research priority number five from Eagles (2014) emphasizes the importance of building public support for PAs. A particular need is to investigate programmes, such as partnerships with gateway communities, to create public support of parks and reciprocally support their natural and cultural resources. Given that many gateway communities worldwide are rural and vulnerable to economic decline, any advancement of a partnership between the town and park may mitigate the external threats and create greater potential and stability for the local citizens.

2.2.2 Gateway communities

The geographic theory postulates that gateway cities offer an entrance and exit to an area, such as a national park, monument or scenic trail (Burghardt, 1971). Gateway communities can contribute to the park as much as the park may contribute to the community, yielding the goal of a sustainable, reciprocal partnership (Wells and Bradon, 1992; Lincoln, 2011). A challenge of this partnership is the capacity needed to balance the low-density usage within the park with the expanding economic development needs of the gateway towns (Stoker *et al.*, 2021). Due to growth and increased visitation, tourism issues could include loss of community character, natural resource degradation and economic inequality (Pretty and Smith, 2004; Park and Pellow, 2011; Slocum and Curtis, 2016).

Gateway communities receive economic benefits from park visitors (Moore and Barthlow, 1998; Stynes and Sun, 2003; Mules, 2005), build public support (Fisher and Durrance, 1972; Walpole and Goodwin, 2001; Ervin *et al.*, 2007; Plummer and Fennell, 2009) and enhance conservation practices (Howe *et al.*, 1997; Brown and Harris, 2005; Judkins, 2015). Howe *et al.* (1997) suggest a conservation benefit by addressing the preservation and access mission of parks and PAs by including the gateway communities within the park's ecological footprint. Since the reciprocal benefits go both ways for gateway communities and the shared park zones, advantages include cooperative management (McMahon, 1999).

Many national parks share a common boundary with gateway communities. For instance, Thompson (2000) acknowledges that a trail will link neighbourhoods to a greenway contributing to a shared vision between both resources. Localities have a vested interest in assuring public access to the healthy ecosystem provided by a trail or park and its surrounding landscape, which protects their water and foodsheds. Partnerships with trailside communities and local citizens can enhance the trail's relevancy, promote connectivity and social support, aligning increased wellness, as well as protect against incompatible development, misuse and environmental degradation.

Despite the many benefits, Kurtz (2010) reported that the era of federal and state public land protection that grew from the mid-1960s to 1980 threatened the economic wellbeing of gateway communities, especially those dependent on resource extraction. Kurtz reports that smaller, rural communities, that could no longer depend on timber or mineral economies when the lands become protected, were potentially harmed. The economic link, frequently touted as a benefit to gateway communities and the neighbouring national park, is further disputed. Aguirre (2006) found that park visitors respected the gateway community, although about half of the visitors made no economic contribution to the community. Yet, the study also recognized the lack of visitor support businesses, as there appear to be gaps in services specifically offered for tourists. Even where businesses benefit from increased tourism, residents may feel stress on current infrastructure, specifically landfill use, water and sewer services, and excess traffic congestion (Dunning, 2005; Joyner *et al.*, 2019). The need for coordination between the gateway town and visitors is evident. This is essential, even more today, as Covid-19 created a surge in outdoor recreation use, increasing management challenges for park managers and communities, impacting park resources and negatively influencing the visitor experience (Derks *et al.*, 2020).

2.2.3 The Appalachian National Scenic Trail

The United States National Park Service (NPS), established in 1916, is a federal agency that plans and manages some of the nation's prized natural, cultural and historic landmarks for future generations. Shortly after the NPS's formation, regional planner Benton MacKaye published 'An Appalachian Trail: A project in regional planning' in the *Journal of the American Institute of Architects* (MacKaye, 1921). Combining social and environmental challenges, his proposal suggested economic redevelopment through a land management strategy for the eastern mountain range of North America. MacKaye describes the 'problem of living' as an economic one, believing that growing industrialization and the extraction of resources in the make-up of our planned landscape was unsustainable – for both the land and the people. Shortly after this proposal, in 1925, MacKaye and other like-minded outdoor enthusiasts established the Appalachian Trail Conference (King, 2000; Anderson, 2002), a not-for-profit organization that mobilized volunteer groups from local hiking clubs and communities to plan and manage the Trail (Appalachian Trail Conservancy, 1981).

His regional planning solution was built on four concepts. First was the creation of a foot trail. Utilizing the many footpaths found along the Appalachian Mountains, it was anticipated that merely linking existing trails could establish a single trail. Recognizing that a 2000-mile trail is formidable in length, MacKaye suggested that this Appalachian Trail length be organized into shorter segments to aid local management. Second, MacKaye proposed constructing shelter camps along the Trail, each about a day's walk apart. Volunteers would be responsible for the construction and maintenance of these structures, often called 'lean-tos'. The third concept was relocating urban populations into small cooperative settlements along the Trail 'for recreation, for recuperation and for study' (MacKaye, 1921, p. 328). This idea was uniquely driven to encourage regional planning for the Trail and still encourage local involvement. Lastly, establishing food and farm camps along the way would give citizens another way to give 'back to the land' (MacKaye, 1921, p. 328). This approach to a project in housing and community architecture would support the temporary volunteer-visitors' need for the necessary food and wood to construct the rural mountains' communities. Locally sourced products would help rural economies remain sustainable.

The nearly 100-year history of the Trail is not without drama. While volunteers completed a continuous path in 1937, some researchers report that the haste in building the connected ribbon of Trail had left MacKaye's full vision of cooperative settlements and camps behind (Foresta, 1987). Lacking was the '[f]orest as well as agricultural land [that] might prove an opportunity for steady employment in the open. But this depends upon a new deal for our agricultural system' (MacKaye, 1921, p. 327). His cooperative 'recreation camps' were a means to social and mental wellbeing and a more equitable economy through land-use planning. Further, the increased use of the Trail during the 1960s threatened the 'casual' management structure of the Trail.

For example, the Trail closures experienced during the 1960s and 70s in northern Virginia were reported in the Trail guide: 'In 1966, however, key property owners withdrew permission for the Trail to cross their land. This necessitated temporary rerouting on Highway 601 ...' (Potomac Appalachian Trail Club, 1970, p. 136). This was eventually resolved by a local club (Potomac Appalachian Trail Club) and Conservancy partnership. From these experiences, the Trail has served as a vision for improving access via private lands, primarily where few other opportunities existed (Helmer *et al.*, 2020). In the Cumberland Valley of Pennsylvania, the Trail was forced to follow roads, and the encroaching urbanization threatened not only the footpath, but agricultural lands and several cultural landscapes (Korostoff *et al.*, 1991). Again, the cooperating efforts of the local trail maintaining group, the Conservancy and government agency partners were necessary to plan and execute relocation onto a protective greenway corridor.

To address this threat, in 1968, the National Trails System Act (NTSA Public Law 90-543), and a follow-up amendment of the Act in 1978, secured the necessary federal funds to establish a permanent route for the Trail that would preserve the character and experience of the Trail greenway. The leading management document, the 'Appalachian Trail Comprehensive Plan' (officially the *Comprehensive plan for the protection, management, development and use of the Appalachian National Scenic Trail*) (Appalachian Trail Conservancy, 1981, preface), defines this character:

> The Appalachian Trail is a way, continuous from Katahdin in Maine to Springer Mountain in Georgia, for travel on foot through the wild, scenic, wooded, pastoral, and culturally significant lands of the Appalachian Mountains. It is a means of sojourning among these lands, such that the visitors may experience them by their own unaided efforts. In practice, the Trail is usually a simple footpath, purposeful in direction and concept, favoring the heights of land, and located for minimum reliance on construction for protecting the resource. The body of the Trail is provided by the lands it traverses, and its soul is the living stewardship of the volunteers and workers of the Appalachian Trail community.

While the Act establishes that the overall responsibility for administering the Appalachian National Scenic Trail rests with the Secretary of the Interior through the NPS, the plan builds on the current abiotic and biotic conditions along the footpath and directs the management of the Trail through a cooperative management system. Because of the unique role of building and managing the Trail with the established 31 trail maintaining clubs, a cooperative agreement between the Appalachian Trail Conservancy and NPS recognizes the Conservancy as a guarantor of specific Trail management responsibilities. The Trail shifted from being an informally managed trail park to being a full partner with the NPS System, offering federal protection and funding (Mittlefehldt, 2010, 2013). The cooperative agreement distinguishes the Conservancy's roles and responsibilities, the substantial involvement of the NPS as the lead agency and liaison to other federal and state agencies, and supports the volunteers of the individual trail clubs as the core of the Trail's cooperative management system.

The partnership is also unique since the federal, state and local agencies whose land the Trail passes through, as well as the local citizen volunteer trail maintaining groups, share the management of the Trail system. Thus, the federal government works directly with the Conservancy who facilitates the day-to-day management by local trail maintaining groups. The Trail's management is complex, covering 14 states, dozens of national and state parks and forests, and a protected corridor (a swath of land averaging about 1000 ft in width) surrounded by more than 250,000 acres making it one of the largest park units in the Eastern United States (National Parks Conservation Association, 2010). This collaborative effort allows thousands of volunteers to take part in the management of the Trail, funds for which would otherwise have to come from federal funds and employee capacity. The Conservancy and its affiliated clubs provide over $3 million annually in contributed volunteer services (National Parks Conservation Association, 2010). This public–private cooperative partnership formalized a long-term relationship that has made the Trail one of the premier hiking experiences available to the public and a model for other long-distance trails worldwide (Burch, 1979; Bristow, 2019). The partnership builds on the efforts of thousands of volunteers, dedicated to preserving the Trail.

To further protect the character and experience in perpetuity, federal funds acquired Trail corridor lands between 1979 and 2005 as part of the NPS land protection programme. The developed corridor was essential to protect not only the Trail but the surrounding natural and cultural landscapes (Leonard, 1979). The iconic Trail symbol (Fig. 2.1) is found at nearly every road crossing, representing the Trail's route through the mountains.

Beyond the monumental undertaking of maintenance for a 2200-mile footpath, the Conservancy is expanding management as envisioned by MacKaye's proposal. The next essential step is to recruit additional local stewards for the day-to-day protection and management of their local Trail section (McMillan, 2008). Fuelled

Fig. 2.1. The iconic marker for the Appalachian Trail (photo courtsey of the authors).

by the management plan, the relationship of gateway communities along the Trail is defined by the Comprehensive Plan (Appalachian Trail Conservancy, 1981) as follows:

- Appalachian Trail Community – A broad term including all those interested in or related to the Appalachian Trail: hikers, volunteers, landowners, federal and state agency personnel, local officials, and citizens of the towns through which the Trail passes (p. 1).
- The stewardship of private landowners and townspeople's involvement along the Trail is an important tradition reflected in the system (p. 5).
- Continuing the tradition of going through several towns that have long been a part of the Appalachian Trail experience (p. 23).

Following the vision of MacKaye (1921), the management plan formally recognizes the importance of the trailside gateway towns with their public officials and citizens, and the importance of private land ownership, as some of the route of the Trail transverses sidewalks and paths in the numerous towns. In doing so, the gateway towns are identified as partners with the Trail.

A membership meeting with agency partners and volunteers was convened in the late 1990s to assess the Conservancy's mission and the acquisition project's success. The meeting culminated in a need for change, taking the organization from the Appalachian Trail Conference to the Appalachian Trail Conservancy. A name change marked a shift in identity, with the Conservancy looking beyond the maintenance and acquisition of the mere footpath to land conservation, stewardship and education. A vehicle for this new vision would be to support and leverage the connections with trailside communities.

To meet these goals, the Conservancy established a partnership programme with the local communities, following MacKaye's 1921 proposal. The Conservancy describes four criteria to become designated as an 'Appalachian Trail (AT) Community' (Appalachian Trail Conservancy, 2012):

- Community support is evidenced by the creation of an advisory committee. Suggested candidates for this committee would be representatives of the county or community (elected officials, etc.), representatives of the Chamber of Commerce, tourism development association, business association or equivalent, local hiking, and AT maintaining club, etc.
- Hosting an annual AT volunteer project, event or celebration. The Conservancy regional offices can provide ideas and suggestions.
- AT related educational or service-learning programme or project. Teacher participation in Conservancy education programmes qualifies.
- Language for the protection of the AT is included in land-use plans, planning tools, ordinances, or guidelines, or there is demonstrated support to amend, change or add such plans.

After pilot programmes were run with two communities in 2006, a series of discussions, vetting and input led to the programme's launch in 2010. Hot Springs, North Carolina, is one of the oldest participants in the AT Community programme. While small in population (577 people, US Census, 2019), this town hosts thousands of hikers annually, as a supply and resting point and is also an attraction in itself for outdoor recreation enthusiasts drawn to the Blue Ridge Mountains in western North Carolina. The Trail runs directly through the centre of the town (Fig. 2.2).

Extensive work has taken place between the Hot Springs Appalachian Trail Advisory Committee and the Conservancy members over the years. It has enabled planners to understand the relationship and benefits of gateway communities and the national park. In a recent survey, 48% of respondents considered economic development and tourism as the Trail's most positive impact; 27% of respondents felt the Trail's contribution to community character as its most positive impact; and 11% of respondents observed the Trail's protection of natural resources and the promotion of environmental quality as its most positive impact (Magee,

Fig. 2.2. Trail route through Hot Springs, North Carolina (photo courtesy of the authors).

2019). Community leaders also describe the desired interest in a stronger connection with Conservancy and Trail management.

The NPS and Conservancy together guide administration through the cooperative management system, including local land management partners and volunteer trail clubs. More than a simple footpath along the Appalachian Mountains, the Trail encompasses an entire social ecosystem, traversing 14 states and numerous habitats. The Conservancy works with partners to protect ecosystem integrity, function and resiliency through improved restoration, connectivity and adaptive management. The linkages between the gateway communities and the Trail have grown over nearly 100 years. What happens in the next 100 years will be served by this study.

In sum, the long-standing partnership between the NPS and Conservancy has now secured a protected route for the Trail. MacKaye's first goal is achieved. Trailside campsites and lean-tos are currently found a day's walk apart; the second goal is achieved. The AT Community is beginning to address MacKaye's third and fourth goals around community groups and food and farm camps. A need to better understand these gateway towns' characteristics, their role in supporting conservation initiatives and growing sustainable outdoor recreation economies is a goal of this chapter.

In particular, since the programme has been in place for ten years, we sought to identify how well-known the programme may be for residents in our gateway towns and perceptions of the programme's benefits to their communities. Second, given the ten-year history of the AT Community programme, we seek to determine if there is an increase in conservation strategies and beneficial economic impacts.

2.3 Methodology

To determine the characteristics of the Trail's gateway towns, two survey instruments were prepared. The original survey was approved by the host University Institutional Research Board to support one of the author's Masters thesis that investigates four of the 47 communities in the designated AT Community programme: Damascus, Virginia; Harpers Ferry-Bolivar, West Virginia; Hot Springs, North Carolina; and Monson, Maine. The second survey was administered to the wider gateway communities along the length of the Trail. The instrument was pretested before targeting the two samples of the gateway towns.

The instrument was administered by the Conservancy to all the designated communities in the programme and neighbouring residents. The instrument remained open from July until September of 2020 to yield the greatest possible sample. We covered the entire length of the Trail with deliberate oversampling from the four original designated communities.

Based on the survey's wide distribution, we cannot determine the response rate. Yet, the comprehensive coverage and oversampling yielded a combined and unique sample size of 177 responses. Given the sampling approach, the results are merely descriptive in this exploratory study of the Trail gateway communities' perception of positive economic development and increased conservation initiatives resulting from the partnership.

2.4 Results

Our results provide an overview of the perceptions of the AT Community programme by gathering basic characteristics of the respondents and assessing the conservation and economic benefits to the community. The following summary and tables provide the reader with the findings from our exploratory study.

For analysis purposes the sample was aggregated by the Conservancy's four management regions: New England (NERO), Mid-Atlantic (MARO), Virginia and West Virginia (VARO), and the Southern states (SORO). Table 2.1 shows the distribution of our sample by these management regions. Almost 90% of the sample came from the Southern states, Virginia and West Virginia or Mid-Atlantic.

To explore characteristics of the sample, the survey inquired about the reasons people move to the community (Table 2.2). Please note it was possible to give multiple reasons, so several answers could be given. In fact, 22 people noted two reasons for their move, and 24 said that four of our examples were the *reason* to move. Interestingly, 13 people were drawn to live in the area following

Table 2.1. Survey distribution, grouped by the four Conservancy management regions.

Conservancy management region	Frequency	%
New England (NERO)	21	11.9
Mid-Atlantic (MARO)	58	32.8
Virginia and West Virginia (VARO)	48	27.1
Southern (SORO)	48	27.1
Unknown	2	1.1

Table 2.2. Why did you move to your community? Select all that apply.

Reason	Total	No answer
Ideal geographic location	55	116
Proximity to my home	47	125
Proximity to the Trail and other recreational amenities	56	116
Unique character	65	107
Experience while thru-hiking the Trail	13	159

Note: n=171

their Trail hike. Living and/or working close to the Trail is seen to be a popular draw.

To further define our sample, we asked about the survey respondents' role in the town. Table 2.3 shows primary roles within the community as residents, employees, business owners, stakeholders or volunteers. About 80% of the survey respondents live and/or work in a designated AT Community. Others were members of the local hiking club.

Table 2.4 describes the participants' length of residency. Most participants lived in or near the AT Community for at least ten years, while about 40%

Table 2.3. Respondent roles.

Role	Frequency	%
Live in Community	54	31.6
Live & work in Community	44	25.7
Work in Community & resident elsewhere	40	23.4
Other connection: retired, club member	33	19.3

Note: n=171

Table 2.4. Respondents' length of residency.

Length of residency	Frequency	%
0–2 years	18	13.3
3–6 years	18	13.3
7–9 years	17	12.6
10+ years, but not originally from the Community	71	52.6
Born & raised	11	8.1

Note: n=135

of participants lived or worked in the Community for less than ten years. The cross-tabulation of years and relationships, summarized in Table 2.5, adds to this interpretation. It is important to note that the participants who live and work in the area are not native-born even though most have been in the area for over ten years, indicating a strong attachment to place near the Trail.

To assess the environmental success of the AT Community programme, the survey questioned participants about the conservation and outdoor recreation strategies their town employed since becoming an AT Community. Table 2.6 summarizes those initiatives that garnered at least 12% (n=14) of the sample. The responses are aggregated by Conservancy management zones to prioritize future needs. Individuals were able to select multiple conservation initiatives.

The development of trail resources in the community not associated with the Trail was the most frequent new action acknowledged by our sample of the strategies. This indicates a broad appeal in expanding community assets for outdoor recreation-related opportunities. In particular, land conservation programmes, like watershed protection, and the creation of farmers' markets were found to be commonly employed, especially in the mid to southern sections of the Trail (i.e., MARO, VARO and SORO). In many cases, respondents articulated that a town may have introduced multiple initiatives. In some cases, respondents noted these practices were in place before the designation as an AT Community. Others reported the need to create additional conservation practices. The reader is also reminded that these same Southern management regions accounted for 85% of the sample size (Table 2.1), so the strong responses also reflect the popularity of the actions in the South.

Table 2.5. Respondents' length of residency vs role.

Role of respondents	0–2 Years	3–6 Years	7–9 Years	10+ Years	Born & raised	Total
Live in Community & work elsewhere	9	9	4	23	5	50
Live & work in Community	4	4	6	11	3	28
Work in Community & resident elsewhere	5	3	5	24	1	38
Other connection: retired, club member	0	2	2	13	2	19
Total	18	18	17	71	5	135

Table 2.6. Since becoming an AT Community, what changes in conservation strategy have you seen?

Conservation strategy	NERO	MARO	VARO	SORO	Total
Creation of farmers' market	0	6	9	10	25
Development of environmental education	1	7	9	7	24
Development of Trail resources in the community excluding the AT Community	6	20	17	15	58
Protection of watersheds	1	14	2	8	25
Protection of forests	1	12	4	7	24

Note: strategies that were promoted by at least 12% of the sample.

Finally, the survey queried the respondents about the perceived economic benefits of the AT Community programme (Table 2.7). The majority believed the programme contributed positively, while only six respondents did not think there was an economic advantage. The comments for the question reported that those who did not work in the community did not observe economic benefits first-hand.

2.5 Discussion

By tabulating the length of residence and role in the community, as well as assessing recent conservation tactics, we identified the importance of the received benefits of Trail resources and the perceived positive economic benefits, as well as the support for increasing conservation landscapes, plans and strategies. The gateway towns, such as those participating in the AT Community programme found along the Appalachian Trail, can continue to benefit from the proximity to a park and PA, and likewise, the park gains local stewards and advocates along the way. This win–win scenario can be expected to grow over time.

Many respondents have been long-time neighbours of the Trail, living and/or working in the area for more than ten years (Table 2.5), and many moved due to geographic location and proximity to recreation assets. This may illustrate a useful finding for Trail advocacy, given

Table 2.7. Awareness of economic benefit resulting from participation in the AT Community programme.

	NERO	MARO	VARO	SORO	Total
Yes	3	24	17	25	69
No	0	3	2	1	6
Maybe/don't know	4	20	19	5	48
Total	7	47	38	31	123

that respondents are fairly evenly split between newcomers and long-time residents. The Conservancy has an opportunity to leverage new residents' likely interest in Trail management as stewards or advocates (McMillan, 2008). This builds on Otero *et al.* (2008) who report that amenity migration occurs where newer citizens are attracted to the resource for the perceived outdoor benefits.

The AT Community programme achieves both short-term and long-term benefits. Designated communities become linked and can exchange ideas with one another as a network to collaborate regionally on planning efforts. For example, one respondent commented, 'It is nice to hear what other communities are doing, hear about struggles they may be having. Learn new ideas such as to how to engage the local community to the AT and the AT Community.' Long-term benefits include expanding park and PA stewardship, increasing landscape-level conservation and improving trailside communities' economic health. Local foodsheds, as inspired by farmers' markets (Table 2.6), found in many gateway towns, support the need anticipated by MacKaye's (1921, p. 328) proposal and fourth goal: the 'establishment of food and farm camps giving citizens another way to give back to the land'. It is clear from the preponderance of interest in Trail-related recreation that since the gateway towns have become designated through the AT Community programme, new trails and trail resources in the region have been built. Anecdotally, many of these new trails are spurs to the Trail or recent rail-to-Trail conversions, expanding access to Trail-related outdoor recreation for residents and tourists alike. Given the growing use of outdoor recreation assets (Stoker *et al.*, 2021), natural resource protection is necessary. It will be important for park managers and communities to build visitor use management strategies focused on stewardship to mitigate resource damage.

This research identifies how a gateway community found along the Trail can use MacKaye's model to build on the local linkages for parks and PAs and compares it to contemporary environmental and economic challenges. Given the Trail's length, varied ecosystems and that multiple economies exist along the route, numerous gateway towns offer the NPS an additional resource to protect this national park.

2.6 Conclusion

MacKaye's vision has been evolving over the past 100 years. Even with the challenges found along the Trail over the last century, the strength of the cooperative model between federal, state, and local agency partners, the Conservancy and the thousands of dedicated volunteers driven by the love of the Trail has made it one of the most resilient parks, running over the spine of the eastern seaboard.

Rural communities have always been attractive destinations for second homes (Stynes, 1997). We found that nearly half of the respondents have lived in the town for less than ten years (Table 2.4). Further, recognizing regional differences along a 2000-mile national park, several key findings of new conservation initiatives, like watershed programmes and the creation of farmers' markets, were referenced as new since the gateway town's formation and recognition. These findings deserve additional examination by in-depth interviews with the partnered members.

Natural resource- and outdoor recreation-based tourism, as found along the Trail, are clear economic development opportunities for gateway towns to parks, forests, reserves and PAs (Cerveny *et al.*, 2020). The next 100 years will bring new questions; some we can anticipate, like climate change and increasing visitor use, others are unknown, like those from the pandemic. Since our trailside towns channel visitors to and from the Trail, the gateway communities are likely to be the first to see changes. As the Conservancy continues its efforts to conserve lands through a broad landscape framework, the next 100 years can offer a more regional planning approach.

In 2021, the Conservancy launched a new strategic plan with interconnected goals, including land protection (building capacity to manage for healthy and resilient Trail lands); cooperative management (strengthening shared stewardship and planning, and perhaps a new vision of the Appalachian Trail for the next century of collaborative management to include gateway communities); engaging new people and new partners; and fostering a culture of inclusion. The Conservancy aspires to cultivate greater strength and resilience as it manages and protects the Trail along with its greater ecosystem of communities and people. The last goal includes a focus on a more robust framework in working with trailside community partners.

The next 100 years of park management will need to consider a socially inclusive strategy that continues to address the challenges of climate, sustainability, and support of authentic cultures (Leung *et al.*, 2018). For this holistic approach, the AT Community programme needs to work cooperatively with Trail partners to be economically viable, support the environment's protection, and provide social and economic equity. The next generation of park stewards are young people. Once we garner their interest and unbridled enthusiasm, the bonds with one another via the connection with the Trail can serve the collective future of Trail lands and people (Judkins and De Camara, 2020). Parks and PAs, such as the Appalachian Trail, can build on this partnership with gateway communities and, united, face the next century of responsible management as envisioned by MacKaye over 100 years ago. In sum: 'National protected areas policy is changing to reflect progressive global discourses on best practices for more inclusive conservation programs, but the place-specific conditions, in which policy is transformed into lived experience, present perhaps the greatest challenge' (Conz, 2014, p. 241).

The authors report no conflict of interest.

References

Aguirre, J.A. (2006) Linking national parks with its gateway communities for tourism development in central america: nindiri, nicaragua, bagazit, costa rica and portobelo, panama. *Pasos. Revista de Turismo y Patrimonio Cultural* 4, 351.

Anderson, L. (2002) *Benton MacKaye: Conservationist, Planner, and Creator of the Appalachian Trail*. Johns Hopkins University Press, Baltimore, MD, USA.

Appalachian Trail Conservancy (ATC) (1981) Comprehensive plan for the protection, management, development and use of the Appalachian National Scenic Trail. Appalachian Trail Conservancy.

Appalachian Trail Conservancy (ATC) (2012) Appalachian Trail Community: a designation program of the Appalachian Trail Conservancy. Available at: https://appalachiantrail.org/wp-content/uploads/2020/02/ATCommunityApplication.pdf (accessed 1 February 2022).

Bello, F.G., Lovelock, B. and Carr, N. (2017) Constraints of community participation in protected area-based tourism planning: the case of Malawi. *Journal of Ecotourism* 16(2), 131–151.

Bristow, R.S. (2019) Altruistic adventure voluntourism: help manage the park you visit. In: Jenkins, I. (ed.) *Adventure Tourism and Outdoor Activities Management: A 21st Century Toolkit*. CAB International, Wallingford, UK, pp. 145–154.

Brown, R. and Harris, G. (2005) Comanagement of wildlife corridors: the case for citizen participation in the Algonquin to Adirondack proposal. *Journal of Environmental Management* 74(2), 97–106. DOI: 10.1016/j.jenvman.2004.08.005.

Burch, W.R. (1979) *Long Distance Trails: the Appalachian Trail as a Guide to Future Research and Management Needs*. School of Forestry and Environmental Studies, Yale University, New Haven, CT, USA.

Bureau of Outdoor Recreation (1966) *Trails for America: Report on the Nationwide Trails Study*. United States Department of Interior, Bureau of Outdoor Recreation, Washington, DC. Available at: https://cdn2.assets-servd.host/material-civet/production/images/documents/trails-for-america-1966.pdf (accessed 1 February 2022).

Burghardt, A.F. (1971) A hypothesis about gateway cities. *Annals of the Association of American Geographers* 61(2), 269–285.

Cerveny, L.K., Sánchez, J.J., Helmer, M. and Milnor, A. (2020) Public lands, tourism, and community connections. In: Selin, S., Cerveny, L.K., Blahna, D.J., and Miller, A.B. (eds) *Igniting Research for Outdoor Recreation: Linking Science, Policy, and Action*. US Department of Agriculture, Forest Service, Pacific Northwest Research Station, Portland, OR, USA, pp. 115–132.

Conz, B.W. (2014) Conservation and Maya autonomy in Guatemala's western highlands. In: Stevens, S. (ed.) *Indigenous Peoples, National Parks, and Protected Areas: A New Paradigm Linking Conservation, Culture, and Rights*. University of Arizona Press, Tucson, AZ, USA, pp. 241–260.

Derks, J., Giessen, L. and Winkel, G. (2020) COVID-19-induced visitor boom reveals the importance of forests as critical infrastructure. *Forest Policy and Economics* 118, 102253. DOI: 10.1016/j.forpol.2020.102253.

Dunning, A.E. (2005) Impacts of transit in national parks and gateway communities. *Transportation Research Record* 1931(1), 129–136.

Eagles, P.F. (2014) Research priorities in park tourism. *Journal of Sustainable Tourism* 22(4), 528–549.

Eagles, P.F. and McCool, S.F. (2002) *Tourism in National Parks and Protected Areas: Planning and Management*. CAB International, Wallingford, UK.

Ervin, J., Spensley, J., Hayman, A., Lopez, C., Blyther, R, *et al.* (2007) *Capacity Action Planning for Protected Areas: A Quick Guide for Practitioners*. The Nature Conservancy, Arlington, Virginia.

Fisher, R.M. and Durrance, D. (1972) *The Appalachian Trail*. National Geographic Society, Evergreen, Colorado.

Foresta, R. (1987) Transformation of the Appalachian Trail. *Geographical Review* 77(1), 76–85.

Foster, C.H. (1987) *The Appalachian National Scenic Trail: A Time to Be Bold*. Appalachian Trail Conservancy.

Helmer, M., Miller, A.B., Barborak, J.R., McCool, S.F. and Leung, Y.F. (2020) Global dimensions: trends, lessons, and collaborative learning. In: Selin, S., Cerveny, L., Blahna, D., and Miller, A. (eds) *Igniting Research for Outdoor Recreation: Linking Science, Policy, and Action*. US Department of Agriculture, Forest Service, Pacific Northwest Research Station, Portland, OR, USA, pp. 133–148.

Howe, J., McMahon, E.T. and Propst, L. (1997) *Balancing Nature and Commerce in Gateway Communities*. Island Press, Washington, DC.

Jones, R., Fly, M., Talley, J. and Cordell, K. (2003) Green migration into rural America: the new frontier of environmentalism? *Society & Natural Resources* 16(3), 221–238.

Joyner, L., Lackey, N.Q. and Bricker, K.S. (2019) Community engagement: an appreciative inquiry case study with Theodore Roosevelt National Park gateway communities. *Sustainability* 11(24), 7147.

Judkins, J. (2015) *Understanding Local Perceptions of Management and Values of Long Distance Trails*. Nicholas School of the Environment, Duke University, Durham, North Carolina, USA.

Judkins, J. and De Camara, C. (2020) The future of conservation: next generation forest ambassadors. *A.T. Journeys, Fall*, 32–34.

King, B. (2000) Trail Years. In: Rubin, R. (ed.) *Special 75th Anniversary Issue, July 2000 – Appalachian Trailway News*, pp. 2–1652-63.

Korostoff, N., Yanher, T. and Johns, T. (1991) The Appalachian Trail across the Cumberland Valley: A case study of linear park design with cultural landscape preservation, agricultural preservation, and landscape ecology. *Proceedings from International Linear Parks Conference: Parkways, Greenways, Riverways: A Partnership for Beauty and Progress*. Appalachian State University, Boone, NC, USA, pp. 33–45.

Kurtz, R.S. (2010) Public lands policy and economic trends in gateway communities. *Review of Policy Research* 27(1), 77–88.

Leonard, R. (1979) Protecting the long trail resource: A problem analysis. In: Burch Jr, W.R. (ed.) *Long distance trails: The Appalachian trail as A guide to future research and management needs. Symposium on the Appalachian Trail*. Yale University, New Haven, CT, USA, 11-13 October, 1977.

Leung, Y.F., Spenceley, A., Hvenegaard, G. and Buckley, R. (2018) *Tourism and Visitor Management in Protected Areas: Guidelines for Sustainability*. IUCN, Gland, Switzerland.

Lincoln, C.A. (2011) Exploring links between gateway communities and national parks: a quantitative assessment. Doctoral dissertation. Oklahoma State University, Tulsa, OK, USA.

MacKaye, B. (1921) An Appalachian Trail: a project in regional planning. *Journal of the American Institute of Architects* 9(10), 325–330.

MacKaye, B. ([1928] 1990) *The new exploration: A philosophy of regional planning*. University of Illinois Press, Urbana-Champaign, IL, USA.

Magee, T. (2019) *Hot Springs, North Carolina: An A.T. community's progression from 2006 to 2019*. Appalachian Trail Conservancy.

McMahon, E. (1999) Gateway communities. *Planning Commissioners Journal* 34, 6–7.

McMillan, L. (2008) Turning mountain migrants into mountain stewards: best practices, future promise in bridging stakeholder divides. In: Moss, L., Glorioso, R. and Karuase, A. (eds) *Understanding and Managing Amenity-led Migration in Mountain Regions*. The Banff Centre, Banff, Alberta, Canada, pp. 273–281.

Mittlefehldt, S. (2010) The people's path: Conflict and cooperation in the acquisition of the Appalachian Trail. *Environmental History* 15(4), 643–669.

Mittlefehldt, S. (2013) *Tangled Roots: The Appalachian Trail and American Environmental Politics*. University of Washington Press, Washington, DC, USA.

Moore, R.L. and Barthlow, K. (1998) *The economic impacts and uses of long-distance trails: Featuring a case study of the Overmountain Victory National Historic Trail*. US Department of the Interior, National Park Service, Washington, DC, USA.

Mules, T. (2005) Economic impacts of national park tourism on gateway communities: the case of Kosciuszko National Park. *Tourism Economics* 11(2), 247–259.

National Parks Conservation Association (2010) *Appalachian National Scenic Trail: A special report*. Center for State of the Parks, National Parks Conservation Association, Washington, DC, USA.

Otero, A., Gallego, E. and Lonac, A. (2008) Case study: Amenity migration – challenge for better understanding gateway communities and national parks to conserve common resources. In: Moss, L., Glorioso, R. and Karuase, A. (eds) *Understanding and Managing Amenity-led Migration in Mountain Regions*. The Banff Centre, Banff, Alberta, Canada, pp. 171–182.

Park, L.S.H. and Pellow, D.N. (2011) The case of the missing mountain: Migration and the power of place. In: Burton, L.M., Kemp, S.P., Leung, M., Matthews, S.A., and Takeuchi, D.T. (eds) *Communities, Neighborhoods, and Health*. Springer, Cham, Switzerland, pp. 111–126.

Plummer, R. and Fennell, D.A. (2009) Managing protected areas for sustainable tourism: prospects for adaptive co-management. *Journal of Sustainable Tourism* 17(2), 149–168.

Potomac Appalachian Trail Club (1970) *Guide to the Appalachian Trail: From the Susquehanna river to the Shenandoah National Park*. Potomac Appalachian Trail Club, Washingon, DC, USA.

President's Commission on Americans Outdoors (US) (1987) *Americans Outdoors: The Legacy, the Challenge, with Case Studies: The Report of the President's Commission*. Island Press, Washington, DC, USA.

Pretty, J. and Smith, D. (2004) Social capital in biodiversity conservation and management. *Conservation Biology* 18(3), 631–638.

Schottanes, J. (2021) *An Assessment of Four Selected Communities Along the Appalachian Trail in Relation to Emile Benton MacKaye's Original Vision of Regional Planning*. Master's thesis, University of Massachusetts, Amherst, MA, USA.

Selin, S., Cerveny, L.K., Blahna, D.J. and Miller, A.B. (2020) *Igniting Research for Outdoor Recreation: Linking Science, Policy, and Action*. U.S. Department of Agriculture, Forest Service, Pacific Northwest Research Station, Portland, OR, USA, p. 257.

Slocum, S.L. and Curtis, K.R. (2016) Assessing sustainable food behaviours of national park visitors: Domestic/on vocation linkages, and their implications for park policies. *Journal of Sustainable Tourism* 24(1), 153–167.

Stevens, S. (2014) *Indigenous Peoples, National Parks, and Protected Areas: A New Paradigm Linking Conservation, Culture, and Rights*. University of Arizona Press, Tucson, AZ, USA, pp. 3–12.

Stoker, P., Rumore, D., Romaniello, L. and Levine, Z. (2021) Planning and development challenges in western gateway communities. *Journal of the American Planning Association* 87(1), 21–33.

Stynes, D.J. (1997) *Seasonal Homes and Natural Resources: Patterns of Use and Impact in Michigan*, Vol. 194. US Department of Agriculture, Forest Service, North Central Forest Experiment Station, St. Paul, MN, USA.

Stynes, D.J. and Sun, Y.Y. (2003) *Economic Impacts of National Park Visitor Spending on Gateway Communities, Systemwide Estimates for 2001*. Final report to National Park Service. Department of Park, Recreation and Tourism Resources, Michigan State University, East Lansing, MI, USA.

Thompson, J.W. (2000) Getting trails on track. *Landscape Architecture* 90(6), 80–133.

US Government Accountability Office (1994) *National Park Service: Activities Outside Park Boundaries Have Caused Damage to Resources and Will Likely Cause More*. United States Government Publishing Office, Washington, DC, USA.

Walpole, M.J. and Goodwin, H.J. (2001) Local attitudes towards conservation and tourism around Komodo National Park, Indonesia. *Environmental Conservation* 28(2), 160–166. DOI: 10.1017/S0376892901000169.

Wells, M. and Bradon, K. (1992) *People and parks: Linking Protected Area Management with Local Communities*. World Bank.

3 Building a Gateway to the Arctic: A Political Economy Perspective on Tourism Development and Conservation in Finnish Lapland

Dorothee Bohn[1] and C. Michael Hall[2*]
[1]Umeå University, Sweden; [2]University of Canterbury, New Zealand

Abstract
Research on gateway communities conventionally focuses on micro-level aspects that emerge in the nexus of environmental conservation, tourism development and local livelihoods. However, the embeddedness of places and the local tourism sector in the modes of production, consumption and capital circulation of contemporary capitalism remains oftentimes unaddressed. This chapter, therefore, adopts a political economy perspective and examines the macro-frameworks that condition tourism development in Finnish Lapland, in tandem with attempts to consolidate its gateway position to the Arctic. The aim is to encourage a more nuanced view on gateway status in regions where tourism development is driven by multi-scalar stakeholder interests and embedded into competitive regional development initiatives. While the devastating effects of climate change and human induced pollution on the fragile terrestrial and maritime Arctic ecosystems are well recognized, the spatial reimagination of Lapland in the Arctic represents another neoliberal step towards the total commodification of the environment.

Keywords: Arctic tourism, Finnish Lapland, regional smart specialization, Arctification, political economy

3.1 Introduction

The notion of a gateway is conventionally employed in tourism studies as a reference to small- and medium-sized towns that provide an entrance point to amenity-rich landscapes, protected areas (PAs) or public lands. Respective research focuses commonly on the micro-level aspects that emerge in the nexus of environmental protection, tourism development and the adjacent community, such as the commercial effects of national park visitation on local economies, amenity migration patterns, the opportunities of tourism as a livelihood for Indigenous populations, residents' attitudes towards tourism development and the manifold conflicts that emerge around nature conservation (Howe *et al.*, 1997; Mules, 2005; Hall, 2015; Panta and Thapa, 2018; Herbert *et al.*, 2020). Bottom-up planning and local business ownership often appear as the prime solutions to foster local agency and to overcome frictions surrounding tourism development and ecological protection (e.g., Frauman and Banks, 2011; Joyner *et al.*, 2019). However, advocacy of

*Corresponding author: michael.hall@canterbury.ac.nz

 Tourism Transformations in Protected Area Gateway Communities (eds S.L. Slocum *et al.*)
DOI: 10.1079/9781789249033.0003

small-scale locally owned tourism, under the umbrella of sustainability, often ignores issues of political economy and the structural role of gateways in exploiting their hinterlands as part of contemporary processes of globalization and neoliberal capitalism (Bianchi, 2018), thereby echoing Britton's (1991, p. 466) sentiment that 'by treating tourism almost solely as a discrete economic subsystem, many revealing links have been missed between tourism and other politically and theoretically important geographic issues which demonstrate the wider role and position of tourism in capitalist accumulation'.

The study by Nelson and Hines (2018) on gentrification through amenity migration to a gateway town in rural USA demonstrates vividly that remote places are bound in a relational sense to the webs of accumulation in global capitalism, given that the migrants and their capital are from elsewhere. Unfortunately, many studies are framed uncritically with respect to the existence of a discrete 'community' as the scale of observation. Cohen and McCarthy (2014) emphasize that this oversimplification is problematic because of the underlying romanticizing of a 'community' as the universally good opposite of an anonymous society and the authoritarian state (Hall, 2008). Such a conception misses not only the power structures and divisions that exist within a population, but also masks how a community and its individuals are embedded into the contemporary politics of neoliberal capitalism (Hall, 2008; Brenner *et al.*, 2010).

In transport and tourism geography, gateways carry a much more comprehensive connotation. As a principal place concept, a gateway refers to an entrance/exit point between territories of dissimilar modes of production or environmental properties (Burghardt, 1971). During the 20th century, at the height of Fordism, gateway cities were theorized as settlements that arise as entries to resource extraction regions and function as service centres for the adjacent tributary area (McKenzie, 1933). Nonetheless, different spatial entities, such as cities, ports and even regions, might operate as a gateway for which a temporal stability is characteristic 'as they commonly emerged at the convergence of inland transport systems and through the long-term accumulation of infrastructure and investments' (Rodrigue *et al.*, 2013, p. 68). A key tenet in that early mobility literature is, therefore, to understand gateways not just in their own right but also their strategic positioning in networks, particularly of resource exploitation (Hall, 2014).

The strategic politico-economic role of gateways was reinforced during the global geo-economic and geopolitical transformations that took place from the 1970s onwards and triggered new geographies of statehood in the form of rescaled national territories (Brenner, 2009). Additional spatial units, such as cross-border regions, supranational conglomerates or city regions, emerged alongside the decentralized nation state in order to accommodate the inherent instability of capitalism and the manifold consequences of globalization (Jessop, 2003). The simultaneous shift of power to subnational levels was a pragmatic move to tackle regionally specific problems that national macroeconomic policies were unable to resolve, while enhancing democracy and the participation of local citizens, as well as a consolidation of neoliberalism (Cohen and McCarthy, 2014). Since the 1990s, these processes have been accompanied in OECD (Organization for Economic Co-operation and Development) countries by a focus on the knowledge-based economy, which is driven by quests for international competitiveness through entrepreneurialism, innovation, learning, bottom-up governance and regional specialization (Brenner, 2009). As such, gateways and their hinterland regions occupy a significant position in political economic terms with respect to generating a competitive edge in a globalized world through connecting and controlling regional, national and international networks that govern flows of goods, people, capital and information (Moisio, 2018). Research on global production networks (GPNs) highlights not only the critical role of gateways for logistics and transport, industrial processing, corporate control, service provision and knowledge generation but also addresses the need to consider these functions in relation to the hinterlands, as gateways might enable strategic coupling or grow to the detriment of their adjacent peripheries (Scholvin *et al.*, 2019).

Tourism is deeply embedded in this interplay between places, economy, transport and information and communications technology (ICT) circuits. Gateway destinations typically dominate the hierarchy of tourism networks in national or regional systems with respect to constituting the first encounter with a destination region that might influence the travellers' whole trip experience (Lew and McKercher, 2002). As such, gateways possess considerable economic and distributive power (Hall, 2015). The centrifugal influence of gateways materializes in the concentration of tourist flows in the vicinity of the arrival cities (Raun *et al.*, 2020). Indeed, as Carson *et al.* (2020) note, a major contributing factor is the prevailing 'growth centre' paradigm in regional planning that spurs development in urbanizing places but rarely leads to spillovers from cities to rural hinterlands where tourism represents a warranted economic opportunity. Yet, a gateway status for international tourism can be a key asset for regions or cities and reinforces capital accumulation and place branding strategies, especially by providing access to polar areas (Hall, 2008, 2014). For example, Herbert *et al.*, 2020 outlined the case of Ushuaia, which became one of the most prosperous places in Argentina through monopolizing its access to Antarctica.

In seeking to better understand polar tourism gateways, this chapter adopts a political economy perspective and examines the wider strategic frameworks that condition tourism development in Finnish Lapland in tandem with the region's attempts to consolidate its Arctic gateway position. The aim is to provide a more nuanced view on gateway status where tourism development is driven by multi-scalar stakeholder interests and embedded into regional development initiatives that seek a competitive position based on innovation, new institutional arrangements and particular segments of the population. These competitiveness-centred policy approaches determine how the gateway governs and legitimizes land use, climate policies, environmental protection and sustainable management of its hinterlands. Finnish Lapland, and the region's strong political ambitions to align itself with the Arctic, which has become an environmental, political and economic hotspot (Roussel and Fossum, 2010), represents an interesting case. On the one hand, the devastating effects of anthropogenic climate change and pollution on the fragile terrestrial and maritime Arctic ecosystems are well recognized. Global climate change is affecting the northern hemisphere's high latitudes at rates double the rest of the earth, a process called Arctic amplification (Hall and Saarinen, 2020). Simultaneously, however, the spatial reimagination of Lapland in the Arctic represents another neoliberal step towards the commodification of the environment.

The chapter first outlines the tenets of political economy and, in particular, conceptualizations of state rescaling and its application in tourism research. Secondly, the evolution of tourism in Finnish Lapland in conjunction with state spatialities is presented before elaborating on the contemporary role of the Arctic wilderness as a resource. Thirdly, the region's pursuits of generating a gateway position with respect to branding and transport are illuminated from a political economy perspective. We conclude by discussing the implications of such an approach for tourism and sustainability and the future north.

3.2 The Political Economy, Spatial Restructuring and Tourism Gateway Nexus

'Political economy' refers to the examination of the relationship between the economic spheres of life, including production, consumption and wealth distribution, and historical, political, biophysical, sociocultural and geographical contexts. The most renowned proponents of political economy are Karl Marx (1818–1883) and Friedrich Engels (1820–1895), who critiqued the work of their forebears and contemporaries on the capacity of the free market and private property to spawn wealth and individual freedom beyond interventionist governments (Milonakis and Fine, 2009). Capitalism, as Marxian political economy underscores, is characterized by an antagonistic, yet exploitative relationship between the ruling bourgeoisie, who own the means of production and dominate the political arena, and the working class, who must sell their labour power. The key to capital accumulation lies, therefore, in the extraction of surplus value generated by the wage-dependent worker who experiences a state of alienation, as she holds neither control of the working process nor the final product. Moreover, the structural growth imperative of capitalism necessitates an ever expanding and diversifying market that leads to a chronic instability of the system and is ultimately built on the exploitation of finite natural resources. Marx (1973) theorized that this phenomenon causes distinct ruptures, called metabolic rifts, in social and ecological systems that are evident today in terms of climate change, pollution and the accelerating loss of biodiversity.

In contrast to political economy, neoclassical economics emerged as an orthodox school of thought that sharply separated the social and historical from the economic (Milonakis and Fine, 2009). The latter became a matter of mathematical modelling that comprehended the market as a 'value-neutral' instrument for the resource allocation of rational and utility-maximizing actors (Clark, 2016). Nevertheless, political economy, with a critical undertone, did not cease but found wide appreciation in the social sciences. Even though there exists philosophical and epistemological disparity between the different schools of political economic thought, a common denominator for contemporary political economy studies is to conceive of capitalism as 'conflictual and unstable, incapable of solving its own internal problems and productive of the very socio-spatial inequalities that its proponents believe it can (at least in principle) overcome' (Sheppard, 2011, p. 320).

For this chapter, the geographical political economy approach to state rescaling, informed by regulation theory, is of particular interest. This mode of inquiry is concerned with how the relationships of public, private and third sectors under capitalism are organized into a regime of accumulation and social regulation in space and time (Sheppard, 2016). Western scholars emphasize that the prevailing nation-state regulation of the economy, coupled with Fordism/Keynesianism, fell into crisis from the 1970s onwards due to the internal contradictions of capitalism and external destabilization by globalization and was superseded by different variants of neoliberalism and decentralized modes of governance (Brenner, 2009).

Characteristics for the latter are the simultaneous upward supra-national expansion of the state (e.g., the European Union, EU), the downwards passing of power to local and regional governments and sideway collaboration via public–private partnerships. These state transformations are best understood as path-dependent but experimental. On the one hand, regulatory landscapes inherited from earlier rounds of statehood determine options for future decision-making while it underscores, on the other, that state transformation processes play out differently – often with unintended side-effects – across space, time and institutional contexts (Brenner *et al.*, 2010). Moreover, Sheppard (2016) adds that the role of the state is not reducible to mere reactionary acts to ensure capital accumulation on behalf of a ruling capitalist class, as critics of rescaling often suggest (see Mykhnenko and Wolff, 2019). On the contrary, democratic states are heterogeneous entities that exhibit a certain degree of autonomy from the economy with respect to the strategic–relational capacity of shaping capitalism and responding to the needs of the citizenry (Sheppard, 2016). A topical example are the far-ranging restrictions in the wake of the Covid-19 pandemic that disrupted capital accumulation globally in all components of the travel and hospitality value chain (Hall *et al.*, 2020).

The knowledge-based economy gained popularity from the 1990s onwards as a promise to overcome the material and environmental limits of capitalism through business solutions and absolute decoupling (Moisio, 2018). In the European context, the knowledge-based economy features most prominently in the place-based development agenda of 'smart specialization' (S3). This policy approach aims at fostering competitiveness through the bottom-up mobilization of territorial assets, technological innovation and entrepreneurial discovery. Establishing an S3 framework has also been a precondition for regions' eligibility to access European structural and investment funds during the 2014–2020 funding period. Tourism is well represented as a strategic S3 priority across Europe, both in regions where tourism is an established sector and in regions where tourism development serves as an economic diversification strategy (Biagi *et al.*, 2021). In terms of the EU's cohesion policy goals, research has shown that a tourism sector, based on high visitor volumes, low-skill employment and little connection to related industries generates no significant value added for erasing regional disparities (Romão and Neuts, 2017). Hall and Williams (2020) remark that tourism contributes best to development and innovation through the creation of linkages to other sectors. In addition to enhancing the value chain of travel and hospitality through cross-sectoral cooperation, tourism functions as a means to support human mobility and as an image-maker for a region. Especially in place branding, which emerged as a cornerstone of territorial governance and spatial competitiveness, tourism is important. The latter shapes and makes new places, in the northern Nordic region in particular, through cross-border collaboration and partnerships between municipalities (Prokkola, 2011). 'Visit Arctic Europe' is an example of such interventions, which targets the creation of the Arctic Europe brand, marketing, packaging products of small tourism firms and the creation of interregional transport solutions (Visitarcticeurope.com, n.d).

These spatial policy changes, that are linked to issues of metagovernance (Amore and Hall, 2016), have significant bearings upon the management of PAs. In the context of the EU, supranational structures plus regional and local authorities have been taking over former state conservation responsibilities accompanied by non-state actors, such as voluntary organizations, NGOs (non-governmental organizations), private enterprises and local civil societies in conservation. The common framing of these shifts, as a more democratic and bottom-up approach to decision-making, frequently overshadows the simultaneous commodification of conservation for tourism that has been promoted as the solution to biodiversity loss (Büscher and Fletcher, 2015). In the case of Finland, governance rescaling and neoliberal marketization of environmental protection co-evolved through public–private partnerships and policies aiming at the generation of financial yield. Apostolopoulou *et al.* (2014) emphasize this context, particularly the intensifying use of national parks for nature-based tourism as a tool to boost regional development. For Büscher and Fletcher (2015), this 'accumulation by conservation', which turns the negative environmental contradictions of contemporary capitalism into a model of accumulation, bears the danger of fostering business-as-usual behaviours instead of triggering far-ranging change that responds to environmental and climate crises. Yet, the tourism literature on gateways has largely neglected these political economy aspects of state restructuring, despite the clear role of gateways as enablers of the exploitation of the resources of the hinterland. This, therefore, refers to both the spatial imaginary of gateways within different forms of networks, the strategic positioning of places as gateways (Hall, 2015) and the entanglement of communities in global circuits of capital accumulation.

3.3 Finnish Lapland – From Nordic to Arctic Tourism

Lapland is the northernmost region of Finland and the most sparsely populated area in Europe with 1.8 people per km^2 totalling 180,000 inhabitants. Sixty per cent of the population lives in the urban centres of Rovaniemi, Kemi and Tornio. The Indigenous Sámi community accounts for around 7000 persons. The region is often labelled 'the last wilderness in Europe' because of its large swathes of uncultivated land, of which around 30% is protected by different environmental legislations (Luke, n.d.). The strict nature reserves, which partially prohibit recreationally motivated entry, leisure and tourism use, are firmly integrated into the management of the seven national parks, while the 12 wilderness areas allow traditional forms of land use, such as reindeer herding, hunting, fishing and also tourism (Saarinen, 2019).

Economically, Lapland is dependent on mining, the processing of wood and steel, and tourism. The economic significance of tourism in Lapland has expanded considerably during the past decade. Before the outbreak of the Covid-19 pandemic in early 2020, about 8000 people, including seasonal staff, were employed in programme services and hospitality (House of Lapland, n.d.a). However, the actual employment effect varies between different municipalities; in western Lapland, for instance, 39 to 46% of all employed persons work in the tourism sector (Satokangas, 2013). Tourism's share of the regional gross domestic product amounted to 5.7%, and estimations suggest that in 2017/2018, 200 million euros were invested in the sector (House of Lapland, n.d.a). The accommodation capacity of Lapland, including official as well as cottage and Airbnb rentals, totals 107,300 beds (House of Lapland, n.d.a).

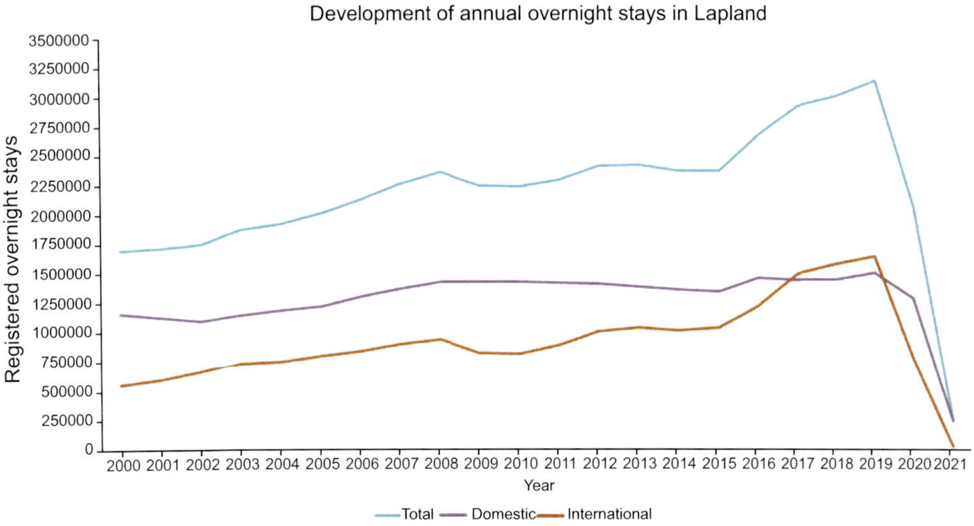

Fig. 3.1. Development of overnight stays in Finnish Lapland (adapted from Tilastokeskus, n.d.).

Fig. 3.1 shows the development of registered overnight stays in Finnish Lapland. In 2017, the registered nights spent by international travellers exceeded, for the first time, domestic visitors, especially the long-haul market from Asia, which has grown since 2015. Lapland has also extended southwards because of the region's international success as a place brand. Kuusamo, which belongs to North Ostrobothnia province, is often marketed as Lapland.

Tourism in Lapland has always been nature-based, yet the utilization of nature, and its incorporation into capital accumulation, has developed along the lines of the region's economy and its governance scales. Organized tourism emerged in the beginning of the 20th century as a state-led initiative. Finland, which gained its independence from Russia in 1917, promoted domestic tourism, especially in Lapland and Karelia, as a means to bring its citizens closer to their fatherland, to spur national patriotism and to assimilate areas that were regarded as culturally non-Finnish (Lähteenmäki and Colpaert, 2020). Tourism development focused particularly on the geopolitically significant Petsamo region, which was ceded from Russia in 1920. Finland not only received access to the Arctic Ocean for sea transportation, but the area proved to be rich in mineral resources; nickel mining became a profitable business (Hautajärvi, 2014). The Finnicization policy of the state encouraged migration to Petsamo, and the newly built road between Rovaniemi and Petsamo allowed tourists to travel up to the North (Lähteenmäki and Colpaert, 2020). In addition, the national government, together with the Finnish Tourist Association, advanced tourism development projects in Lapland. Modern functionalist hotels, which followed the international chic of that time, plus simpler tourist cabins and hostels that were designed to convey neo-national romanticism, were placed in the newly established nature preserve areas so that bourgeois travellers could gaze at the landscape (Hautajärvi, 2014).

Most of Lapland's touristic infrastructure was destroyed during the Second World War, and it was not until the 1960s that tourism recommenced, yet under a different pretence. Tourism development in Lapland was no longer in the hands of central government, state-owned subsidiaries, such as the alcohol monopoly Alko, and the Finnish Tourist Association, but steered by increasingly self-governing municipalities and private entrepreneurs (Hautajärvi, 2014). The purpose of tourism shifted from the state's nation-building prestige project to a replacement industry for the rapidly diminishing primary sector in Finnish peripheries. Municipalities in Lapland faced rural decline through outmigration at the time, and mass tourism skiing centres promised to partly absorb these structural changes. As wealth, leisure time and car ownership had increased significantly within the Nordic welfare state model, tourism was accessible to a broader spectrum of Finnish society. The general trend in Nordic countries towards second homes and outdoor activities, based on Everyman's Rights, also led to a boom in cottage construction in Lapland. With downhill skiing's development as a mainstream leisure activity, around 20 winter sport resorts, of which Levi, Ylläs, Saariselkä and Pyhä-Luosto are the largest today, were constructed in the vicinity of the region's seven national parks and 12 wilderness areas. Capital accumulation through tourism had profound effects on nature conservation. Lapland's 'landscape was turned into a raw material' (Hautajärvi, 2014, p. 432), with a prop-like character for tourism development. This unfolded both in the rapid expansion of touristic infrastructure and in the representation of nature.

The internationalization of tourism received a major boost in the mid-1980s with the strategic shift towards promoting Lapland as the official home of Santa Claus and the building of the Santa Claus Village in Rovaniemi (Pretes, 1995). This site advanced to a key attraction of the region and Christmas charter tourism, mainly for a British market. Within the global experience economy trend, the region became more attractive for central European travellers due to the development of soft nature-based winter activities, such as snowmobile safaris, husky sleigh rides, reindeer farm visits and aurora borealis hunting tours (Rantala *et al.*, 2018).

As a by-product of Finland's upwards governance rescaling that came along with the country's EU membership in 1995, the Regional Council of Lapland was founded (Arter, 2001). A key objective of the council has been regional development through lobbying for the social and economic interests of Lapland, planning, as well as granting national and EU funding to projects that are in line with the funding period's strategic targets (Lapin Liitto, n.d.). These alterations in the political landscape had profound impacts on tourism development. For instance, regional strategic tourism planning, steered by the council, emerged as an amendment to municipal spatial tourism planning in the 1990s. The economic growth of tourism, in addition to enhancing the connectivity via road, train and the region's five airports, has been the main strategic aim ever since, and nature represents the main touristic asset for the region. Furthermore, entrepreneurship has been established as the favoured form of tourism development. The EU's regional development efforts, such as S3, influenced tourism policy in Lapland mainly through requirements for accessing public funding. Together with the bioeconomy – primarily forestry, mining and metal industries – tourism is now a key sector in the region's S3 specialization, which is called 'Arctic Smartness' (Arcticsmartness.eu, n.d.).

The Arctic is another spatial scale that has been incorporated into the political economy of Lapland and its tourism sector. The Arctic, as a conglomerate of eight nation states (A-eight), comprising Finland, Sweden, Iceland, Canada, the USA, Russia, Norway and the Kingdom of Denmark, was established in 1996 under the Arctic Council. This intergovernmental forum also includes Indigenous peoples' representatives and the scientific community, and aims at stimulating cooperation across the eight member countries (Nord, 2016). Neither the private sector nor society in Lapland identified with the Arctic prior to Finland's membership in the Council, although there existed political and scientific interest in the region concerning security and transborder efforts to tackle environmental pollution (Keskitalo, 2004). Over the years, the Arctic advanced to an international economic and political hotspot with respect to climate change and access to the region's natural resources. Moreover, Arctic tourism has been growing significantly, which refers not only to the physical visitation of the circumpolar North but also to a shift in Nordic countries towards 'Arctification', which denotes the rise in tourism experiences that are built around the aurora borealis, coldness, snow, nature and Indigenous culture (Müller and Viken, 2017). These products employ the Arctic, instead of the Fennoscandinavian North, as a reference frame and target primarily international and increasingly high-end consumers. The underlying marketing imaginary promotes pastel-coloured scenery, aurora night skies and landscapes that seem to be 'empty' (Saarinen, 2019) as luxury goods.

Unlike other Nordic regions, Arctification in Lapland is not only limited to the vacation industry but also materializes in the region's political attempts to secure an Arctic gateway position through place branding and promotion. Nature and tourism are central components in this pursuit. The former is the main resource for economic development, while the latter serves as an image-maker for the region.

3.4 The Political Economy of Lapland's Arctic Gateway Position – Putting Tourism and Nature to Work

Establishing a gateway status in the Arctic has been of growing importance for Lapland's political economy. This is particularly evident in the area's current regional development programme, its S3 framework and regional marketing. Lapland's provincial programme 2018–2021 is guided by the vision that 'in 2021 Lapland will be Arctic, open and smart. We produce sustainable success in the world's cleanest region' (Hyry *et al.*, 2017, p. 10, authors' translation). The strategic aims for developing the region focus on strengthening the Arctic economy, fostering a borderless environment for work and skill development, creating wellbeing through clean nature, a good living environment, culture and efficient services, plus enabling growth and competitiveness through good accessibility (Hyry *et al.*, 2017). The aspired Arctic gateway status becomes evident in sentences like 'Lapland connects the Arctic region to the EU' (Hyry *et al.*, 2017, p. 17). Likewise, the region's 'Arctic Smartness' S3 programme seeks to spur the 'sustainable utilization and commercialization of Arctic natural resources and conditions' based on the idea that:

> Lapland is a unique arctic [*sic*] region that strives to be the most innovative and business-driven sparsely populated area in the EU and Circumpolar Arctic by 2022. Arcticness is a natural part of everyday life in Lapland. The people living in Lapland have adapted themselves and their activities to their environment and surrounding nature with a goal of becoming high level experts in Finland and in the EU on sustainable development of the northernmost regions. The Arctic is one of the cleanest and best-preserved places on earth but facing many dynamic and complex changes. In addition to challenges, global changes bring great potential. Lapland, as the northernmost region of Finland and European Union, is in the middle of these universal changes and opportunities.
>
> (Arcticsmartness.eu, n.d.)

These excerpts illustrate the strategic spatial realignment of Lapland with the Arctic in order to gain competitiveness in the global market and to access natural resources that are located beyond the region's borders. Hence, the initial focus of Arctic governance, dominated by security and environmental protection (Keskitalo, 2004), has shifted towards a greater pronunciation of the economic dimension of the region (Nilsson, 2018). The neoliberal underpinning of the knowledge-based economy materializes in the S3 strategy, especially regarding the emphasis on well-adapted and entrepreneurial individuals and on Lapland's provincial programme 2018–2021 with respect to the borderless flow of capital and people (Hyry *et al.*, 2017). Yet, nature and unprocessed natural resources represent a major economic pillar for the region. The threat of climate change and pollution-induced environmental degradation to the Arctic flora and fauna are acknowledged but are portrayed in the above-mentioned quote as an opportunity to generate economic profits. Hickel and Kallis (2020) refute this green growth theory by asserting that absolute decoupling from resource use is not achievable against the backdrop of continued growth, and maintaining business-as-usual approaches to cut carbon emissions will not suffice to avoid global warming, with similar criticisms being made of tourism (Hall and Wood, 2021).

The crucial role of tourism as a direct means of capital accumulation, based on the Arctic environment, is highlighted throughout Lapland's regional development programme and should be supported by 'innovative land-use and opportunity creating zoning' (Hyry *et al.*, 2017, p. 23, authors' translation). In tourism marketing efforts, led by Lapland's regional umbrella brand organization, Northern Finland is introduced as:

> Europe's only wilderness, the gateway to true Arctic adventures. What is wilderness anyway? Wilderness is wild and uncultivated land – the way the Earth looks without human intervention. In Finnish Lapland, the vast majority of space is wilderness, tracts of land and water where forests grow free, rivers rush and gurgle. Here, many wild animals are born, live and die without ever seeing a human being.
> (House of Lapland, n.d.b)

Such sentiments misrepresent the actual state of most wilderness areas, which have been altered by human uses and administrative actions for centuries (Saarinen, 2016). Furthermore, instead of a people-free plain, Lapland's wilderness areas are a cultural and economic resource, especially for reindeer herding and Sámi culture. Saarinen (2019) found that wilderness managers clearly favour local traditional land uses over externally driven and highly-commercialized Arctic adventure tourism, including snowmobile tours and dog sledding, especially because the latter rests upon different values and power structures. As the tourism marketing quote above indicates, touristic wilderness emanates from the ideal of an unoccupied hinterland that is waiting to be discovered by the explorer tourist who, nevertheless, requires the assistance and infrastructure of the global travel industry. Such imaginaries have often led to dispossessing local populations of their access to land (Reimerson, 2016). Besides, under the neoliberal scenario of 'accumulation by conservation', which is currently being promoted in northern Finland, the private sector is able to take advantage of environmental preservation through 'responsible' conduct instead of regulation (Büscher and Fletcher, 2015). Such a situation predicates the development of infrastructure and facilities for mass tourism, which is a precondition for Lapland's economy given its trajectory, and commodifies the national parks for tourism rather than as valued conservation spaces facing rapid environmental change. Lapland may be promoted as Arctic, but the rigours of Arctic travel are being softened for visiting tourists, even in the parks (Rantala *et al.*, 2018).

International tourism is also the main vehicle for disseminating the Lapland brand, of which 'the Arctic is an integral part' (Hyry *et al.*, 2017, p. 16, author's translation). The constructed advantage of branding and storytelling has become central to the construction of Lapland's knowledge-based economy (Hall, 2014). Tourism's strong imaginary of a pure and pristine natural Arctic has spilled over to other sectors, such as the bioeconomy, by facilitating a positive product connotation in international markets (House of Lapland, n.d.a). The favourable associations invoked by tourism function as a way to spur capital investments in, and the social acceptance of, other sectors, particularly mining and industry, which generate higher returns to the region.

> Lapland is a multisectoral region where businesses from tourism, bioeconomy, industry to Arctic research all fit in and prosper. Both larger and smaller companies and investors find efficient networks and true partners in the North. Lapland is a pacemaker in tourism and a well-known brand internationally. The region has some of Finland's biggest and most famous destinations and their growth offers many possibilities for new businesses and investors. Lapland has a strong foothold in the industrial sector and is an export region. Processing metals and wood are large export fields and there's high-class know-how in the industrial circular economy. Lapland is a center point for mining and quarrying. Both industry and mining create demand and opportunities for numerous subcontracting businesses.
> (House of Lapland, n.d.a)

This excerpt suggests that the vast space of the Arctic provides enough room for conflict-free operations of all industries. The typical marketing imaginary of Arctic tourism in Lapland, that is also employed for regional and economic promotion, therefore, ignores the impacts of industry on the fragile Arctic environments (see www.lapland.fi/business/).

The aspired gateway position of Lapland to the Arctic manifests in issues related to international connectivity as well. The Regional Council introduces Lapland on its website as:

> The most accessible Arctic region. Compared with other Arctic regions, Lapland is exceptionally accessible in both the digital sense and by air, rail and road. There are five airports in Lapland. The flight time from Helsinki to Rovaniemi is about 1 hour 15 minutes. You can also take a comfortable journey by night train from Southern Finland to Lapland.
> (Lapin Liitto, n.d.)

In order to enlarge the transport network on a global scale, Finland together with Norway initiated a prestige project for constructing a railway line to the Arctic shipping ports in 2017. The route would enhance Lapland's logistic position 'from the periphery to the hub of international transport and transit. The Arctic Railway, which planned to pass through Lapland, would connect Europe to the Arctic Ocean and the Northeast Passage by rail' (House of Lapland, n.d.c, p. 3). This project is advertised by the regional administration as a tremendous economic opportunity for evolving into a transit centre for oil, gas products, wood, minerals and Norwegian salmon. However, a joint evaluation by Norwegian and Finnish bodies, led by Finland's Ministry of Transport and Communications, concluded in early 2019 that the project is economically unviable given the estimated costs of 2.9 billion euros (House of Lapland, n.d.c). In addition, there persists strong social opposition against the rail development. Protests against the construction plans stem mainly from the Sámi community, supported by Greenpeace and Indigenous activists, because the railway would enhance the destruction of Lapland's wilderness areas and accelerate the fragmentation of reindeer pastures (Quinn, 2019). Despite these objections, the regional council decided to maintain the railway project and included it in the land use plan of Northern Lapland, published in 2020. Although there is a lack of national government support for realizing the Arctic Railway, several private investors have expressed interest. According to Peter Vesterbacka, entrepreneur and co-founder of the Finest Bay Area Development, 'previous development plans for the long-discussed project have not considered tourism. He claims that passenger traffic will play an important role in the Arctic Railway, particularly considering that direct flights from Asia will be landing in Rovaniemi' (House of Lapland, n.d.c, p. 8). Again, tourism provides a means to facilitate the gateway position of Lapland and boost capital accumulation through natural resource use and commercial exploitation.

3.5 Conclusions

The case of Finnish Lapland illustrates that a gateway position entails more than the physical location in the vicinity of PAs, and the sustainable use of these environments is not solely contingent on local communities' conduct. From a political economy perspective, tourism development and natural resource utilization are deeply intertwined in the global circuits and flows of capital accumulation. Tourism in Lapland has evolved along the trajectories of the state-economy nexus. In particular, state rescaling practices have had a direct bearing on how tourism has developed. At present, tourism is a major economic pillar for the region, supported by growth-focused strategic planning, and is well integrated into the S3 programme. Within this facet of the knowledge-based economy, competitiveness is achieved through entrepreneurship and the mobilization of the territorial assets of the gateway hinterland. The tourism resource of the region – the Arctic 'wilderness' landscape, charismatic species (reindeer and boreal forest), and Sámi and northern cultures, along with the related constructed advantages of the northern stories of Santa Claus and Arcticness – are put to work in order to generate exchange values built into touristic production and consumption. Tourism marketing in Lapland, therefore, fosters an image of a gateway that makes the Arctic hinterlands, and the resources within them, accessible not only to people but also to capital. From such a perspective, 'conservation' therefore must be profitable in order to be sustainable, creating the present-day inherent contradictions between the region being affected by the climate crisis while simultaneously pursuing growth strategies that exacerbate that same crisis.

The commodification of the Arctic hinterland is an integral part of Lapland's provincial strategy. By promoting a gateway position to the Arctic, Lapland seeks to turn its peripheral location into a central entry point to the region and enable access to natural resources. Tourism has been instrumental in this endeavour by disseminating Lapland's brand image of a clean and unspoiled blank space, in which different industries can prosper side by side. In addition to the brand image, Lapland seeks to consolidate this gateway position by extending transport networks, in particular through the creation of the Arctic Railway. The route would connect Europe to the Arctic Ocean and offer new possibilities for shipping natural resources, especially in a post-glacial or post-Arctic world

(Varnajot and Saarinen, 2021). Yet, the project is not supported by national-level bodies due to a cost–benefit mismatch, and the Sámi community opposes the venture because of its grave environmental and cultural impacts. Instead, interest in realizing the Arctic Railway stems currently from the private sector and international investors, who underscore that tourism would be a means to make the connection profitable.

Finally, the political approach to Lapland and the Arctic as 'natural capital', coupled with the promotion of the hinterland wilderness as no-man's land, raises the question of long-term sustainability and the merits of neoliberal resource governance. The tourism marketing image of a clean and unspoiled pleasure periphery, with luxury services, stands in stark contrast to the daily news depictions of a globe that is battered by environmental destruction and the catastrophic effects of climate change, with the Arctic being among the most affected areas (Hall and Saarinen, 2020). Yet, business-as-usual behaviour only accelerates the disappearance of the pristine Arctic, and one can only wonder what the new hinterland beyond the Arctic Circle will be like (Varnajot and Saarinen, 2021). New post-ecotourism products and luxury will need to be developed for the rapidly changing Arctic landscape, which will have much less snow cover than at present. Undoubtedly, if current trends continue, new nature-based tourism stories will be told on a pauperized landscape by supposedly responsible businesses and governance constellations that seek a balance in the application of the different levels of sustainability. And, tourism researchers will likely still be uncritically extolling the virtues of communities and tourism as a force for conservation and sustainability, without considering their embedded relationship in the circuits and flows of global capital and the emissions, people and destruction of nature that accompanies them.

This work was supported by the Formas funded research project: Climate Change and the Double Amplifications of Arctic Tourism: Challenges and Potential Solutions for Tourism and Sustainable Development in an Arctic Context.

References

Amore, A. and Hall, C.M. (2016) From governance to meta-governance in tourism? Re-incorporating politics, interests and values in the analysis of tourism governance. *Tourism Recreation Research* 41(2), 109–122. DOI: 10.1080/02508281.2016.1151162.

Apostolopoulou, E., Bormpoudakis, D., Paloniemi, R., Cent, J., Grodzińska-Jurczak, M. *et al.* (2014) Governance rescaling and the neoliberalization of nature: the case of biodiversity conservation in four EU countries. *International Journal of Sustainable Development & World Ecology* 21(6), 481–494. DOI: 10.1080/13504509.2014.979904.

Arcticsmartness.eu (n.d.) Arctic Smartness. Available at: https://arcticsmartness.eu/ (accessed 2 February 2022).

Arter, D. (2001) Regionalization in the European peripheries: the cases of northern Norway and Finnish Lapland. *Regional & Federal Studies* 11(2), 94–114. DOI: 10.1080/714004693.

Biagi, B., Brandano, M.G. and Ortega-Argiles, R. (2021) Smart specialisation and tourism: understanding the priority choices in EU regions. *Socio-Economic Planning Sciences* 74(8), 100883. DOI: 10.1016/j.seps.2020.100883.

Bianchi, R.V. (2018) The political economy of tourism development: a critical review. *Annals of Tourism Research* 70(September 2017), 88–102. DOI: 10.1016/j.annals.2017.08.005.

Brenner, N. (2009) Open questions on state rescaling. *Cambridge Journal of Regions, Economy and Society* 2(1), 123–139. DOI: 10.1093/cjres/rsp002.

Brenner, N., Peck, J. and Theodore, N. (2010) After Neoliberalization? *Globalizations* 7(3), 327–345. DOI: 10.1080/14747731003669669.

Britton, S.J. (1991) Tourism, capital, and place: towards a critical geography of tourism. *Environment and Planning D: Society and Space* 9(4), 451–478. DOI: 10.1068/d090451.

Burghardt, A.F. (1971) A hypothesis about gateway cities. *Annals of the Association of American Geographers* 61(2), 269–285. DOI: 10.1111/j.1467-8306.1971.tb00782.x.

Büscher, B. and Fletcher, R. (2015) Accumulation by conservation. *New Political Economy* 20(2), 273–298. DOI: 10.1080/13563467.2014.923824.

Carson, D.A., Åberg, K.G. and Prideaux, B. (2020) Cities of the North: Gateways, competitors or regional markets for hinterland tourism destinations? In: Lundmark, L., Carson, D.B., and Eimermann, M. (eds) *Dipping in to the North: Living, Working and Traveling in Sparsely Populated Areas*. Palgrave Macmillan, London, UK, pp. 285–310.

Clark, B. (2016) *Political Economy. A Comparative Approach*, 3rd edn. ABC-Clio, Santa Barbara, CA, USA.

Cohen, A. and McCarthy, J. (2014) Reviewing rescaling: strengthening the case for environmental considerations. *Progress in Human Geography* 39(1), 3–25.

Frauman, E. and Banks, S. (2011) Gateway community resident perceptions of tourism development: incorporating importance-performance analysis into a limits of acceptable change framework. *Tourism Management* 32(1), 128–140. DOI: 10.1016/j.tourman.2010.01.013.

Hall, C.M. (2008) *Tourism planning*, 2nd edn. Pearson, London, UK.

Hall, C.M. (2014) Will climate change kill Santa Claus? Climate change and high-latitude Christmas place branding. *Scandinavian Journal of Hospitality and Tourism* 14(1), 23–40. DOI: 10.1080/15022250.2014.886101.

Hall, C.M. (2015) Polar gateways: Approaches, issues and review. *The Polar Journal* 5(2), 257–277. DOI: 10.1080/2154896X.2015.1080511.

Hall, C.M. and Saarinen, J. (2020) 20 Years of Nordic climate change crisis and tourism research: a review and future research agenda. *Scandinavian Journal of Hospitality and Tourism* 21(1), 102–110. DOI: 10.1080/15022250.2020.1823248.

Hall, C.M. and Williams, A.M. (2020) *Tourism and Innovation*, 2nd edn. Routledge, London, UK.

Hall, C.M. and Wood, K.J. (2021) Demarketing tourism for sustainability: Degrowing tourism or moving the deckchairs on the Titanic? *Sustainability* 13(3), 1585. DOI: 10.3390/su13031585.

Hall, C.M., Scott, D. and Gössling, S. (2020) Pandemics, transformations and tourism: be careful what you wish for. *Tourism Geographies* 22(3), 577–598. DOI: 10.1080/14616688.2020.1759131.

Hautajärvi, H. (2014) *Autiotuvista lomakaupunkeihin: Lapin matkailun arkkitehtuurihistoria* [From wilderness huts to holiday towns: The architectural history of tourism in Lapland]. Aalto-yliopisto julkaisusarja (2014). Unigrafia, Finland.

Herbert, A., Liggett, D. and Frame, B. (2020) Polarising Ushuaia: informal settlements and tourism in an Antarctic gateway city. *Local Environment* 25(11–12), 849–871. DOI: 10.1080/13549839.2020.1843420.

Hickel, J. and Kallis, G. (2020) Is green growth possible? *New Political Economy* 25(4), 469–486. DOI: 10.1080/13563467.2019.1598964.

House of Lapland (n.d.a) *Business*. Available at: https://www.lapland.fi/business (accessed 2 February 2022).

House of Lapland (n.d.b) *Lapland - Europe's Last Wilderness*. Available at: https://www.lapland.fi/visit/about-lapland/lapland-europes-last-wilderness/ (accessed 2 February 2022).

House of Lapland (n.d.c) *The Arctic Railway– Risks and Opportunities*. Available at: https://www.lapland.f i/uploads/2019/10/c9834acc-arctic-railway.pdf (accessed 2 February 2022).

Howe, J., McMahon, E.T. and Props, L. (1997) *Balancing nature and commerce in gateway communities*. Island Press, Washington, DC, USA.

Hyry, M., Nikander, M., Lintula, J., Kenttäkumpu, J., Jokikaarre, P., *et al.* (2017) *Lappi-sopimus. Lapin maakuntaohjelma 2018-2021 [Lapland contract. Lapland's regional programme 2018-2021*. Available at: http://www.lappi.fi/c/document_library/get_file?folderId=3589265&name=DLFE-32814.pdf (accessed 2 February 2022).

Jessop, B. (2003) The political economy of scale and the construction of micro-regions. In: Söderbaum, F. and Shaw, T.M. (eds) *Theories of New Regionalism*. Palgrave Macmillan, Basingstoke, UK, pp. 179–196.

Joyner, L., Lackey, N.Q. and Bricker, K.S. (2019) Community engagement: an appreciative inquiry case study with Theodore Roosevelt National Park gateway communities. *Sustainability* 11(24), 7147. DOI: 10.3390/su11247147.

Keskitalo, E.C.H. (2004) *Negotiating the Arctic. The Construction of an International Region*. Routledge, London, UK.

Lähteenmäki, M. and Colpaert, A. (2020) Memory politics in transition: Nostalgia tours and gilded memories of Petsamo. *Matkailututkimus* 16(1), 8–34. DOI: 10.33351/mt.85341.

Lapin Liitto (n.d.) *Information about Lapland*. Available at: https://www.lapinliitto.fi/en/information/general-information-about-lapland/ (accessed 2 February 2022).

Lew, A.A. and McKercher, B. (2002) Trip destinations, gateways and itineraries: the example of Hong Kong. *Tourism Management* 23(6), 609–621. DOI: 10.1016/S0261-5177(02)00026-2.

Luke (n.d.) Suojelualueet, talousmetsien monimuotoisuuden suojelukohteet ja luontoarvojen suojelua tukevat tukevat alueet muuttujina vuosi, maakunta, suojeluluokka, maaluokka ja pinta-ala ja osuus [Protected areas, commercial forests' biodiversity protected areas and areas for the protection of natural values as variables year, region, protection category, land category and area and proportion]. Available at: https://statdb.luke.fi/PXWeb/pxweb/fi/LUKE/LUKE__04%20Metsa__02%20Rakenne%20ja%20tuotanto__04%20Metsien%20suojelu/02_metsien-suojelu.px/table/tableViewLayout1/ (accessed 2 February 2022).

Marx, K. (1973) *Grundrisse*. Penguin Books, London, UK.

McKenzie, R.D. (1933) *The Metropolitan Community*. Russell & Russell, Clayton, GA, USA.

Milonakis, D. and Fine, B. (2009) *From Political Economy to Economics: Method, the Social and the Historical in the Evolution of Economic Theory*. Routledge, London, UK.

Moisio, S. (2018) *Geopolitics of the Knowledge-based Economy*. Routledge, London, UK.

Mules, T. (2005) Economic impacts of national park tourism on gateway communities: the case of Kosciuszko National Park. *Tourism Economics* 11(2), 247–259. DOI: 10.5367/0000000054183513.

Müller, D.K. and Viken, A. (2017) Indigenous tourism in the Arctic. In: Viken, A. and Müller, D.K. (eds) *Tourism and Indigeneity in the Arctic*. Channel View Publications, Bristol, UK, pp. 3–15.

Mykhnenko, V. and Wolff, M. (2019) State rescaling and economic convergence. *Regional Studies* 53(4), 462–477. DOI: 10.1080/00343404.2018.1476754.

Nelson, P.B. and Hines, J.D. (2018) Rural gentrification and networks of capital accumulation—A case study of Jackson, Wyoming. *Environment and Planning A: Economy and Space* 50(7), 1473–1495. DOI: 10.1177/0308518X18778595.

Nilsson, A.E. (2018) Creating a safe operating space for business: the changing role of Arctic governance. In: Wormbs, N. (ed.) *Competing Arctic Futures. Historical and Contemporary Perspectives*. Palgrave Macmillan, Cham, Switzerland, pp. 117–138.

Nord, D.C. (2016) *The Changing Arctic*. Palgrave Macmillan, Basingstoke, UK.

Panta, S.K. and Thapa, B. (2018) Entrepreneurship and women's empowerment in gateway communities of Bardia National Park, Nepal. *Journal of Ecotourism* 17(1), 20–42. DOI: 10.1080/14724049.2017.1299743.

Pretes, M. (1995) Postmodern tourism: The Santa Claus Industry. *Annals of Tourism Research* 22(1), 1–15.

Prokkola, E.-K. (2011) Regionalization, tourism development and partnership: The European Union's North Calotte Sub-programme of INTERREG III A North. *Tourism Geographies* 13(4), 507–530. DOI: 10.1080/14616688.2011.570371.

Quinn, E. (2019, September 22) *The Arctic railway. Building a future ... or destroying a culture?* CBC News. Available at: https://newsinteractives.cbc.ca/longform/the-arctic-railway (accessed 2 February 2022).

Rantala, O., Hallikainen, V., Ilola, H. and Tuulentie, S. (2018) The softening of adventure tourism. *Scandinavian Journal of Hospitality and Tourism* 18(4), 343–361. DOI: 10.1080/15022250.2018.1522725.

Raun, J., Shoval, N. and Tiru, M. (2020) Gateways for intra-national tourism flows: measured using two types of tracking technologies. *International Journal of Tourism Cities* 6(2), 261–278. DOI: 10.1108/IJTC-08-2019-0123.

Reimerson, E. (2016) Sami space for agency in the management of the Laponia World Heritage site. *Local Environment* 21(7), 808–826. DOI: 10.1080/13549839.2015.1032230.

Rodrigue, J.P., Comtois, C. and Slack, B. (2013) *The Geography of Transport Systems*, 3rd edn. Routledge, London, UK.

Romão, J. and Neuts, B. (2017) Territorial capital, smart tourism specialization and sustainable regional development: experiences from Europe. *Habitat International* 68(2–3), 64–74. DOI: 10.1016/j.habitatint.2017.04.006.

Roussel, S. and Fossum, J.E. (2010) The Arctic is hot again in America and Europe: introduction to part I. *International Journal* 65(4), 799–808.

Saarinen, J. (2016) Wilderness use, conservation and tourism: what do we protect and for and from whom? *Tourism Geographies* 18(1), 1–8. DOI: 10.1080/14616688.2015.1116599.

Saarinen, J. (2019) What are wilderness areas for? Tourism and political ecologies of wilderness uses and management in the Anthropocene. *Journal of Sustainable Tourism* 27(4), 472–487. DOI: 10.1080/09669582.2018.1456543.

Satokangas, P. (2013) Matkailulla maakunta menestyy – Matkailun tulo ja työllisyysvaikutukset [The region thrives on tourism - Income from tourism and employment effects], 12 lappilaisessa kunnassa vuonna 2011. Multidimensional Tourism Institute, University of Lapland, Finland.

Scholvin, S., Breul, M. and Diez, J.R. (2019) Revisiting gateway cities: connecting hubs in global networks to their hinterlands. *Urban Geography* 40(9), 1291–1309. DOI: 10.1080/02723638.2019.1585137.

Sheppard, E. (2011) Geographical political economy. *Journal of Economic Geography* 11(2), 319–331. DOI: 10.1093/jeg/lbq049.

Sheppard, E. (2016) *Limits to Globalization. Disruptive Geographies of Capitalist Development*. Oxford University Press, Oxford, UK.

Tilastokeskus (n.d.) *Yearly nights spent and arrivals by country of residence by region, type of accommodation establishment, country, information and year*. Available at: https://pxnet2.stat.fi/PXWeb/pxweb/en/StatFin/StatFin__lii__matk/statfin_matk_pxt_11j1.px/table/tableViewLayout1/ (accessed 2 February 2022).

Varnajot, A. and Saarinen, J. (2021) 'After glaciers?' Towards post-Arctic tourism. *Annals of Tourism Research* 91(3), 103205. DOI: 10.1016/j.annals.2021.103205.

Visitarcticeurope.com (n.d.) *Visit Arctic Europe II*. Available at: https://visitarcticeurope.com/ (accessed 2 February 2022).

4 Gateway Community Relationships with A US National Park: A Qualitative Exploration of Community Member, Concessioner and Grand Teton National Park Staff Perspectives

N. Qwynne Lackey[1]*, Kelly S. Bricker[2] and Russell M. Hicks[3]
[1]State University of New York System, New York, USA; [2]Arizona State University, Phoenix, Arizona, USA; [3]Kansas State University, Manhattan, Kansas, USA

Abstract

In the mid-1900s, park and protected area (PPA) management agencies began to acknowledge the environmental, socioeconomic and cultural connections that existed between PPAs and gateway communities. The purpose of this qualitative case study was to understand the effects of shifts in the perspectives of PPA managers, concessioners and gateway community members who work and live in the vicinity of Grand Teton National Park (GTNP). We conducted interviews with 21 park staff, concessions staff and gateway community leaders. Data revealed 15 themes organized into four categories: (a) types of relationships that exist between the community and park; (b) characteristics that describe these relationships; (c) foundations for collaboration; and (d) barriers to collaboration. This research provides insight into the effects of PPA management approaches that are alternative relative to traditional PPA strategies. Additionally, we provide a broader conversation on the state of PPA policy regarding engagement and collaboration with gateway communities.

Keywords: case study, collaboration, destination management, sustainability, landscape conservation

4.1 Introduction

Consistent with contemporary western views on wilderness, the first national parks established in the late 1800s and early 1900s in the United States were managed with the primary goal of preservation (Schelhas, 2009). Great strides were taken to separate the human element from the parks to preserve what European Americans perceived to be their 'natural' and 'untamed' character (Schelhas, 2009). As a result, these parks, and those established worldwide in the decades that followed, were treated as 'islands' and managed in isolation without much consideration for, or input from, the public (Mose and Weixlbaumer, 2007; Keiter, 2013; Dinica, 2016). Most management activities ended abruptly at the political boundaries of parks, and little attention was paid to how external forces, such as local communities, affected national parks (Dinica, 2016; Slocum, 2017). At the

*Corresponding author: qwynne.lackey@cortland.edu

 Tourism Transformations in Protected Area Gateway Communities (eds S.L. Slocum *et al.*)
DOI: 10.1079/9781789249033.0004

same time, little thought was given to how parks impacted the lands and communities outside their designated borders.

However, in the mid- to late-1900s, park and protected area (PPA) managers and stakeholders began to acknowledge the environmental, economic and cultural connections between national parks and the adjacent lands and communities (Keiter, 2013; Slocum, 2017). Human activity and encroachment, that was unimaginable when the first national parks were established, began occurring beyond park borders (National Park Service, 2014). Preventing and mitigating the negative impacts of these activities resulted in the development and demand for solutions that are collaborative in nature (National Park Service, 2014). Thus, PPA management agencies, including the US National Park Service (Park Service), have begun a marked shift away from managing PPAs in isolation, increasing their involvement in partnerships and large landscape-scale conservation initiatives (Keiter, 2013; National Park Service, 2014; Dinica, 2016; Slocum, 2017; Zajchowski *et al.*, 2019). For example, in 2012, the Park Service established an interdisciplinary group, the Connected Conservation Community, to share best practices and provide educational opportunities aimed at improving collaborative stewardship and landscape scale work (National Park Service, 2020a). In addition, the Park Service published a call to action for stewardship and engagement prior to its centennial celebration (National Park Service, 2015). Within this larger call, the Park Service made a goal to 'scale up' its participation in large landscape-level conservation (National Park Service, 2015).

Due to this shift, the relationship between the Park Service and gateway communities is receiving increased attention. Gateway communities – also referred to as adjacent, park-proximal and natural amenity communities – are communities that reside close to or share a border with PPAs. Because of this proximity, gateway communities have a greater likelihood of impacting park resources than communities located farther away. Likewise, many gateway communities benefit from parks thanks to increased wellbeing and economic stability, which are typically attributed to tourism development and other ecosystem services (e.g., Eagles and McCool, 2002; Kerstetter and Bricker, 2012; Bricker and Black, 2015). Alternatively, gateway communities are subjected to the challenges of national park tourism development and management policies. These can include increased population growth and infrastructure stress (e.g., National Park Service, 1998; Acquah *et al.*, 2017), cultural and livelihood shifts (e.g., National Park Service, 1998; Mbaiwa and Stronza, 2010) and social justice concerns, such as displacement or inequitable benefit distribution (e.g., Matose, 1994; Puhakka *et al.*, 2014; Ramos and Prideaux, 2014). Several planning challenges specific to gateway communities in the Western United States have been identified (Stoker *et al.*, 2020). For example, the 'small town' values many gateway community members appreciate are now challenged by increased development and affordability issues (Stoker *et al.*, 2020). These challenges and relationships within the park–gateway community context are dynamic and complex, making gateway communities a particularly unique and interesting PPA community to examine.

There is limited research on this new and emerging area of study, specifically investigating the relationship between US national parks and gateway communities (Lackey and Bricker, 2020). Most studies have examined how gateway community members perceive national parks or the impacts of national parks on their community (e.g., Morris and McBeth, 2003; Bergstrom, 2018; Bergstrom and Harrington, 2018). Others examined dynamics within gateway communities, exploring the priorities of different community segments (Gregor and Koithan-Louderbck, 1997; Bruyere *et al.*, 2002). Additionally, some have studied the impact of specific national park policies on gateway communities (Dunning, 2005; Marquit and Mace, 2015; Gabe, 2016). And more recently, researchers have begun to address the impact of increased desirability as a place to live and work, in addition to visitation (Stoker *et al.*, 2020).

While these studies have increased our understanding of national park gateway communities in the United States, there are several questions that arise. Specifically, questions about the nature of these relationships – especially regarding the forms of collaboration that might exist between national parks and gateway communities – have not been addressed. Therefore, the purpose of this qualitative case study was to understand the relationship between a gateway community and one US national park, Grand Teton National Park (GTNP), from the perspective of Park Service staff, concessioners and gateway community members.

4.2 Methods

We performed a descriptive case study methodology (Creswell and Poth, 2016; Schwandt and Gates, 2017), taking an in vivo approach to using existing theory in qualitative research (Andersen and Kragh, 2010). Because we were interested in understanding relationships and exchanges that occurred between the park and community, we employed social exchange theory as a framework for developing our interview questions and analysing our data. This allowed us to focus in on existing relationships and interactions.

4.2.1 Study setting

The setting for our case study was GTNP and the surrounding community. The park was established to protect its rich natural, geologic and cultural history, and it is located in western Wyoming and on the ancestral lands of the Shoshone, Bannock, Blackfoot, Crow, Flathead, Gros Ventre, and Nez Perce Peoples (National Park Service, 2019b). In 2019, GTNP received more than 3.4 million recreation visits, making it the eighth most visited US national park that year (National Park Service, 2020b). Many of these visitors were served by one of the 23 authorized concessioners or another commercial service provider operating under a commercial use authorization (CUA) (National Park Service, 2019a). A concessioner is a private business that provides appropriate and necessary visitor services within the boundaries of a national park under a contractual agreement. A CUA provider operates appropriate, small-scale visitor services in national parks. Concessioners and CUA providers are separate from the Park Service, but they do play an important role in providing visitor services and often interact with local communities. Therefore, they are of interest in this study.

GTNP is located in Teton County, WY. The county seat of Teton County is the town of Jackson, which is generally considered the primary gateway community. However, during the course of this study, we learned that the park works just as much, if not more frequently, with the leadership of Teton County as it does with the town. Therefore, we broadened our view to examine the relationship between the park and the town and county, which we conceptualized as the 'gateway community' of GTNP.

4.2.2 Data collection

The data for this study were collected as part of a larger study on sustainability in and around GTNP. We conducted semi-structured interviews with Park Service staff, concessioner or CUA staff and gateway community leaders who were recruited using snowball sampling techniques. Interviews were completed in person or by phone and recorded with participants' permission, using a protocol approved by the University of Utah's Institutional Review Board and GTNP. In addition to questions relevant to the overall study, each participant was asked to respond to the following questions: (1) what is the current relationship between the park and community and (2) in your experience, how, if at all, has this relationship changed over the years? All interviews were transcribed using the assistance of Dragon NaturallySpeaking (Home edition, version 15.3) and de-identified.

4.2.3 Data analysis

We analysed participant responses to the two interview questions listed in the previous subsection using an inductive and iterative data analysis process (Miles *et al.*, 2014). First, we independently analysed the data and developed an initial set of codes and themes. Next, we met to compare our findings and engaged in an iterative process until we established an agreed upon set of categories, themes and definitions. Then, we worked independently to compare our results to the original transcripts to ensure our findings were reflected in the data. Lastly, member checking was completed by sharing the findings with participants, and their responses were used to enrich the findings (Rose and Johnson, 2020).

Table 4.1. Emergent categories, themes and subthemes by participant categories.

Category	Theme	Park Service staff (n=4)	Concessions and CUA staff (n=10)	Community members (n=7)
Types of relationships	*Economic*	4	10	6
	Environmental	4	7	6
	Cultural	4	7	6
	Destination management	4	9	6
Characteristics of the relationships	*Positively viewed*	2	8	5
	Increasing collaboration	4	7	5
	Conflict	1	5	5
Foundations for collaboration	*Shared vision*	3	1	3
	Necessary partnerships	2	2	5
	Shared human capital	4	7	5
	Leadership	1	3	3
	Concessioners	1	6	3
	Corporate sponsorship	2	1	3
Barriers to collaboration	*Resource barriers*	2	8	3
	Bureaucracy	0	6	4
Opportunities for increased collaboration	*Transportation*	0	2	4
	Affordable housing	0	2	0
	Renewable energy	1	1	1
	Waste management	0	2	1
	Volunteering	0	1	0
	Destination management	1	3	6

A total of 21 participants completed interviews, which took, on average, one hour to finish (Table 4.1). Our data analysis process resulted in 15 themes organized into four categories that descried the: (a) types of relationships that existed between the community and park; (b) characteristics that described this relationship; (c) foundations for collaboration; and (d) barriers to collaboration.

4.3 Results

4.3.1 Types of relationships

4.3.1.1 Economic

The park has strong economic ties to both the town and county. First, when accounting for the concessioners' employees, the park is one of the largest employers in the county. In addition, the park is a primary attraction for tourists, which are described as the life blood of local businesses. Park Service staff and concessions employees typically shop at local stores, and many park concessioners also source local products to support their operations, creating further economic benefits for the community. Many concessioners operating in the park also choose to support individuals and organizations in the community through scholarships, grants, fundraisers and in-kind donations. Likewise, members of the community frequently support projects and operations in the park through financial donations to the Park Service or to the GTNP Foundation.

4.3.1.2 Environmental

Generally, participants recognize that the park and community share an ecosystem. Many participants described actions that individuals or groups have taken to protect nature resources beyond their respective jurisdictional boundaries. Community residents and organizations will often volunteer to assist with trail maintenance or litter clean-ups inside the park. Likewise, concessions staff members will frequently volunteer to support environmental sustainability efforts beyond park borders, such as removing barbed wire from wildlife corridors. Some participants described how concessioners, in their efforts to enhance their sustainability performance by buying environmentally friendly products from local suppliers, create a demand for environmentally sustainable products, making these products more accessible to others located in the town and county. Residents, organizations and businesses in the town and county have also been strong political supporters of environmental protection in and around the park. For example, many local businesses advocated in support of designating the Snake River Headwaters as a 'Wild and Scenic River', which enhances water and wildlife resource protection in the park.

4.3.1.3 Cultural

The park and community also share cultural ties that enhance quality of life for individuals affiliated with each entity. For instance, the park is an important recreational resource used by many local community members, and access to this resource plays an important role in attracting many community members to the area. Similarly, many Park Service and concession employees, who came to the area seeking employment in the park, have frequently chosen to stay in the community after they stop working inside park boundaries. Some of these individuals have become entrepreneurs or leaders in the local community. This strengthens cultural ties between the park, town and county, which individuals from all three participant pools described as a single, united community. They see each other as neighbours and friends, with whom they have developed connections beyond their professional lives.

4.3.1.4 Destination management

Some participants also recognize that the park and community are part of one tourism system. They acknowledge – either implicitly or explicitly – that both entities serve the same visitors and share a transportation and waste management network. The Park Service has official agreements with Teton county to establish concurrent law enforcement jurisdictions and share other emergency services, such as emergency medical services and structural fire units. Additionally, individuals representing the community, Park Service and concessioners will frequently attend the meetings of, or even serve on, the same local boards, many of which are relevant to tourism destination planning and management. Lastly, some individuals affiliated with the park (primarily concessioners) are also engaged in an initiative to certify Jackson Hole as a sustainable destination.

4.3.2 Characteristics of the relationships

4.3.2.1 Positively viewed

Generally, most participants described the relationship between the park and community as 'good'. Furthermore, many, especially those who worked or lived in the area for less than 20 years, believed that this relationship had been positive and stable for many years. As one concessions employee explained, 'I didn't have to walk into a situation where I had to build [a good relationship]. It was already built, and I just had to maintain and continue it.'

Some participants acknowledged, however, that the positive character of this relationship has improved. For example, two participants discuss how relationships were tense between the park and community decades ago or even before their lifetime. They attributed this to community discontentment with Park Service actions, such as expanding the park border and increasing regulation on certain types of recreation, such as snowmobiling or motor boating. While some segments of the community may still be unhappy with the Park Service because of these issues, these participants explained how these conflicts are largely an issue of the past due to the passing of time, turnover in the community, and a local cultural and economic shift away from agriculture.

4.3.2.2 Increasing collaboration

A few participants described how the amount of collaboration between community members, concessioners and the Park Service has increased on a long- and short-term scale. One Park Service employee attributes this to a general shift favouring increased collaboration with gateway communities that has, and is, taking place within the Park Service. One participant said:

> I think there is history within the Park Service, particularly big western parks, as being kingdoms unto themselves and not having much collaboration with outside communities, even gateway communities. I think there have been some instances of bad blood between communities and parks because they just had no communication ... I think that has dramatically changed in recent history ... I think almost universally, parks are working with gateway communities and seeing that opportunities for sustainability and smart growth have to be a collaborative effort. The Park Service has done better at looking at things from an ecosystem scale and not just, 'Hey, our border stops here. That's it. We are not worried about anything outside of that.' So, I think Park Service wide, there definitely is a broader view than there has been in the past.
>
> (Park Service employee)

This shift in philosophy has also taken place within some concession operations and facets of the town and county, leading to a relatively recent increase in collaborative projects and initiatives. One example was the initiative to certify Jackson Hole as a sustainable tourism destination. There has also been extensive collaboration in the area of waste management. For example, many discuss a partnership between a local farm and a concessioner operating large facilities in the park. In an effort to reduce waste, this concessioner partnered with the local farm, which accepts and composts the concessioner's food waste. The farm uses this compost to grow vegetables, some of which are purchased by the concessioner. This creates a sustainable cycle that has local environmental, economic and cultural benefits. However, the most common example of collaboration is the Zero Landfill Initiative (ZLI). The ZLI is a partnership between Subaru of America, the National Parks Conservation Association and the Park Service. The purpose of the project was to pilot waste reduction strategies in three national parks, including GTNP. The pilot project in GTNP includes many partners, including several concessioners, Teton County Integrated Solid Waste and Recycling and Jackson Hole Public Art. Through this project, connections are forged or strengthened between like-minded individuals affiliated with the Park Service, concessioners and Teton County. Participants from all three participant pools agree that this project united their respective entities, creating one larger community working toward the same goals.

> The Subaru ZLI has brought together people I never would have been involved with. [The category I concessions staff] and I are on the phone regularly, like speed dial almost, and that never would have happened before. Park Service staff were sitting in on the [Integrated Solid Waste and Recycling] board meetings down in Teton County, so the Park had a relationship with Teton County, but the concessioners didn't have a relationship with them other than using them as a facility to get rid of stuff (*sic*) ... Now, all of us can pick up the phone and call Teton County [Integrated Solid Waste and Recycling staff] and say, 'Hey, why is this happening?' I attribute a lot of our working relationship on the ZLI. We have an amazing partnership throughout, and it is because of that programme.
>
> (Park Service employee)

In addition to the direct outcomes, there have been indirect implications. First, relationships are often maintained after the conclusion of a project. This creates a foundation for future, and potentially more complex, collaborative efforts. Second, these projects serve as a testing ground for innovative sustainability solutions that could inspire or inform future projects. For example, lessons learned during ZLI are being used to inform the development of a county-wide food waste composting programme through Teton County Integrated Solid Waste and Recycling.

> The really exceptional part about all of this is that Teton County wants to launch a commercial food waste programme next summer, and they have tons and tons of information about how they can educate people, and businesses that are participating. They got to test out and get a much better sense of equipment and the contamination issues [*sic*], so they have a lot of good intel that will make,

> hopefully, some of the missteps less when they launch their programme.
>
> (Concessions employee)

4.3.2.3 Conflict

Some participants did acknowledge conflict or tension in the relationship between the park and community. Conflict seems to occur when there are conflicting interests or when members or groups within the community are dissatisfied with the actions the other party takes (or fails to take) within their respective jurisdiction. A common example of conflict involves the transportation system. Some participants noted that the community is frustrated with the Park Service for failing to pave or consistently snowplough a section of road, which is part of a popular tourism and commuting route, within their jurisdiction. Participants also comment on how some community members remain unhappy with the restrictions the Park Service places on certain types of recreation in the park. Other conflicts revolve around development. Examples include the Park Service's recent decision to allow more communication towers in the park, and a concessioner's requested amendment to use allowances that would have permitted the concessioner to use their facilities outside the park for housing, warehousing and administration to support their in-park operations. This request was recently denied by Teton County Board of County Commissioners, who ruled that these facilities should be accommodated within the park and not on other land in Teton County. Additionally, there is a sense that some community members and organizations are frustrated by the Park Service's limited engagement with the community beyond the area of waste management. To some degree, participants feel like the community is more willing to work with the Park Service than the Park Service is to work with the community.

> I feel the town of Jackson reaches out often to the Park Service to be good partners and to collaborate with the Park Service. [The Park Service] pretty much comes back ultimately saying 'no'. You know, 'This is how we do business. Thanks for trying to work with us but this is going to be the way'. There's not a lot of flexibility with the Park Service.
>
> (Concessions employee)

4.3.3 Foundations for collaboration

Many participants stated that the relationship between the park and its gateway communities is very different from what they experienced elsewhere. Despite some existing conflict and tension, these participants feel that the relationship between GTNP and gateway communities is more positive than most. Based on interview data, there are many contributing factors that created a strong foundation for this generally positive relationship.

4.3.3.1 Shared vision

While variation exists, participants believed that most residents have a strong environmental ethic. One of the primary reasons many people choose to live in the community is to enjoy the area's rustic character and abundant natural resources. Therefore, many community members and organizations are predisposed to supporting issues, such as natural resource preservation and sustainability. This makes it easier for the Park Service and community to develop shared goals at Grand Teton compared to places where the dominant interests of the community misalign with the Park Service's mission: 'We work as a community. Now, everybody may have different goals perhaps within that community, but the long-term vision I think is there because it is such a special place' (Park Service employee).

4.3.3.2 Necessary partnerships

Furthermore, participants noted that the park and community often need to work together to achieve their shared or respective goals. While jurisdictional boundaries are clear, there is some recognition that the park and community are part of one destination; they serve the same visitors, manage and enjoy the same wildlife, utilize the same waste management system, etc. Segments within each party recognize that they cannot fully achieve success by working within silos.

> When I say 'we' I am talking collectively, these communities, these parks, the National Park Service, the concessionaires, and interest groups. What do we want to see as the future of these parks and what is acceptable in terms

> of a visitor experience and what is acceptable in terms of preserving and protecting the resources of these parks and wildlife? So, being very decisive about that, and then making that happen, not just letting it happen to us or to the parks.
>
> (Community member)

4.3.3.3 *Shared human capital*

The community and park also share a substantial amount of social capital. As mentioned, many participants explained how Park Service employees, concessions employees and community members see each other as neighbours and friends. Members from the park and community work hard to develop and maintain their shared human capital. For example, some Park Service employees and concessions employees invest a lot of time attending meetings and serving on boards at the town, county and even the state level. Likewise, many members of the community make similar efforts to open and maintain lines of communication with representatives from the park. There was also an effort on both sides to stay open, amicable and assume good intentions, an approach which some believed plays a role in developing and maintaining positive relationships. '[Jackson] is our community. It is our churches, it is our schools, it's our local and county government, it's a library, etc. So, it's the community that we belong to or are part of' [*sic*](Park Service employee).

4.3.3.4 *Leadership*

Throughout this chapter, we have mostly discussed the relationship as if it occurs between organizations and communities at large. However, it is important to note that relationships also occur between, and depend on, the actions and choices of individuals. Indeed, during interviews, some participants described the efforts of a few individuals, or 'champions', affiliated with the park or community that play a critical role in developing cross-boundary relationships. These individuals have typically lived and worked in the area for a long time. They also invest a lot of time, sometimes over a period of many years, in engaging with others outside their respective jurisdictional boundaries. While this time invested is often included as part of their professional responsibilities, they frequently choose to exceed their professional requirements. Thanks to these persistent efforts, these champions are able to develop the connections and reputation needed to foster cross-boundary understanding and collaboration in some areas with success.

> You need somebody to be a champion, but you also need someone to be a champion who has leverage and ideally having leverage, like one of these large government agencies or one of the largest businesses, taking that initiative would have certainly made that different [*sic*]. [One leader driving collaborative efforts] was the former head of [an established community organization]. Had [they] not been in that role before, I think it would have been next to impossible to make the progress that [they] made. [They] knew so many people, and [they] had a good reputation and good clout in the community to start to bring this together.
>
> (Concession staff member)

4.3.3.5 *Concessioners*

In addition, GTNP has more approved concessioners than most national parks. Many concessioners play an important role in facilitating the various economic, environmental and cultural relationships between the park and community. In fact, the concessioners are often able to collaborate and contribute more to community projects than the Park Service because they are required to comply with fewer regulations and policies. Additionally, though cross-boundary relations may be important to many Park Service employees and the agency as a whole, they often do not have enough resources, particularly time and money, to expand their collaborative efforts. Some concessioners had more resources, and more flexibility, to dedicate those resources to community engagement initiatives.

> The Park Service is very good at messaging. I think we are the ones that can actually implement it. I think we can implement it in a way where we're not tied in politically, the way the Park Service is with being an arm of government [*sic*] and I think we are able to be more nimble with our responses because we're able to say and talk about the way we like to improve things, how we like to do things, how we would like to change things.
>
> (Concession staff member)

4.3.3.6 Corporate sponsorship

Lastly, the corporate sponsorship of the ZLI was described by many participants as a 'critical engager', which makes this innovative, collaborative partnership possible. Food waste composting in the park and county was a project several partners wanted to pursue for many years. However, the financial resources and expertise needed to develop the necessary infrastructure, and the systems to support that effort, were not available. Corporate sponsorship greatly decreased those barriers, making the project possible: '[The ZLI] would not have happened without Subaru's funding' (Community member).

4.3.4 Barriers to collaboration

4.3.4.1 Resource barriers

When asked about barriers to collaboration between the Park and community, almost every participant mentioned some type of resource barrier. The most common resource barriers were time and money. As alluded, establishing and maintaining healthy cross-boundary relationships requires substantial time and money. This can be a difficult barrier to overcome, especially for the Park Service, which has faced budgetary stagnation or decline in recent years.

> Everybody from myself, to human resources, to our general manager spends a lot of time in each one of our departments helping out ... We have to be there to help them, so it really comes down to time. Maybe if we had extra folks on board, we can make that happen, but is that feasible [*sic*] Is that the right way to go about it? I'm not sure.
>
> (Concession staff member)

4.3.4.2 Bureaucracy

Another barrier can be described as a bureaucratic barrier. The Park Service must comply with the same regulations and policies that govern all national park units nationwide. Some participants representing the community or concessioners see these top-down polices as inflexible and cumbersome barriers to strengthening relationships and collaboration with the Park Service at the local level.

> It's surprising we don't sit down with [the Park Service] more often, and I think we probably could. It shows that they so strongly have their own kind of structure that we can't always just call them up and have a meeting and get things to change because it comes way down from the top on the rules and procedures that they have to follow.
>
> (Community member)

Likewise, frequent turnover within the federal government and Park Service makes it difficult to maintain collaborative relationships long-term. Together, these bureaucratic barriers make it difficult to put the philosophical shift toward collaboration into action.

> There's a growing appreciation for the need to work together rather than to try and do everything by ourselves. What we are struggling with, perhaps, is that we don't know what that looks like, so we are trying to figure it out. We are trying to figure out how we can collectively work together in a way that is going to make it easier for both of us to achieve our respective goals, but there's not a formal mechanism, track record, or what have you [*sic*].
>
> (Community member)

4.3.5 Opportunities for increased collaboration

When asked, all participants believed there are opportunities to increase collaboration, and some were able to identify specific opportunities. Many commented on transportation and affordable housing, and a few described potential renewable energy projects and volunteer-based partnerships. Likewise, some participants hoped that collaboration in the area of waste management would continue and expand. Lastly, some participants described opportunities that generally fall into the category of destination management. Through increased communication, coordination and resource sharing, these participants believed that the park and community would be better equipped to find innovative solutions to the issues they all face.

> More than anything right now, we need everybody to come together. We have all these

> individual entities working at their own silos. They are spending a huge amount of time and money, and they don't have to. If they can work together, they could pool together resources. Spend a lot less money and do a whole lot more. That's not happening except for in the area of waste.
>
> (Community member)

4.4 Discussion

Our study highlights the multifaceted relationship – which includes tangible and intangible economic, environmental and cultural dimensions – that exists between GTNP, the town of Jackson, and others in Teton County. As observed in other PPA-gateway community contexts (see Eagles and McCool, 2002; Keiter, 2013; Bricker and Black, 2015), our results demonstrate how this relationship can benefit and burden both parties. From an economic perspective, tourism in PPAs, and the associated economic value, is an opportunity for PPAs and gateway communities. Visitor entry and tourism operator fees often provide the significant amount of financial capital needed to manage and protect PPAs (e.g., Bushell and Bricker, 2017; Epler Wood, 2017). Visitation is also a source of economic development for gateway communities, directly and indirectly creating sources of income and opportunities to diversify and stabilize local economies (e.g., Eagles and McCool, 2002; Bricker and Black, 2015; Headwaters Economics, 2018). This relationship and its benefits are certainly acknowledged by stakeholders in this study.

GTNP and the community of Jackson are also linked environmentally, as the political boundaries drawn between them have little to no effect on the movement of air, water, flora and fauna between these adjacent spaces. While research has shown that the environmental relationship between PPAs and gateway communities is bi-directional, the nature of this relationship varies substantially between contexts. In this case, many community members benefited from, and valued, various ecosystem services provided by the park, which generally has been associated with positive community attitudes and support toward the park and conservation (Tampakis *et al.*, 2014; Black and Cobbinah, 2018). However, similar to previous studies, negative community attitudes toward the park are based on restrictions the park places on the community (Lindsey *et al.*, 2011; Gandiwa *et al.*, 2013).

Research addressing benefit sharing related to communities adjacent to PPAs has identified the need for a socio-ecological systems approach to understanding the complexities of these relationships (Snyman and Bricker, 2019). Our findings support this need, as we found characteristics of the relationships that support or challenge these various types of collaborations or conflicts. For example, beyond the thread of defining economic, cultural and environmental relationships, we began to understand how these relationships have an impact on increasing collaborative opportunities for sustainability and viewing the park as an ecosystem, rather than a 'kingdom' all to its own. This speaks to the importance of stakeholder engagement. As Snyman and Bricker (2019) note, 'an understanding of local community dynamics, hierarchies, and power-sharing are also essential to develop methods to ensure that the most marginalized and most impoverished are included in benefit-sharing mechanisms' (p. 709). There is a recognition of the mutual economic benefits that support community and park needs through scholarships, contributions to the local economy (i.e., employment, support of local goods and services, fundraising), and mutual environmental benefits through volunteer initiatives within, and on, the periphery of the park boundary, such as environmental advocacy and waste reduction. These efforts are designed around specific needs of the community and park and have led to increased local empowerment when it comes to political issues arising in the region.

While many aspects of these relationships reflect positive collaboration, there are issues and challenges that create some conflicts. The conflicts or disgruntlements with the park are focused on lack of responsibility for issues within either the community or park jurisdiction, often as a result of a lack of resources on behalf of the park. In addition, due to federal regulations and policies, there are limitations on how far community collaboration could feasibly go within the federal framework of the Park Service. While the national park system is a federal entity, guided by federal policies, the fact that gateway communities are often the most affected due to their proximity means innovation in policy development may be necessary to address more regional impacts of parks on communities and communities on parks.

We learned from stakeholders in this study what facilitates and deters from collaboration and perhaps benefit sharing more universally. Overall, there is a sense that the relationships that exist between GTNP and communities in the Jackson area are somewhat unique in the United States. While this study did not address the degree to which the foundations of collaboration were felt or prioritized, it is clear that participants in this study see that a shared goal or vision, partnerships, shared social capital, inspired individuals or champions, private concessions within parks, and corporate sponsorship all greatly enhance, create and often facilitate collaborations between the park and the community. Yet, stakeholders also recognize some limitations. From limited human, time and funding resources to the inflexibility of federal–local policies, several barriers appeared to get in the way of strengthening collaborative efforts. These challenges are not new and require innovation in policy and community engagement strategy, such as this community and park did with their ZLI initiative.

4.5 Limitations and Future Directions

Although saturation was reached, these findings reflect the experiences of a sample of purposefully selected individuals. Therefore, these results may not generalize to all who live or work in or near GTNP. Likewise, this study was conducted within one PPA gateway community context. The nuances of the economic, environmental and cultural relationship likely vary between other PPAs and communities. Furthermore, data collection for this study was completed prior to the start of the Covid-19 pandemic. During the member checking process, several participants noted that the relationship between the park and community has evolved during the pandemic, leading to more collaboration in some instances and increased negative visitor impacts in others. Additional research is needed to understand the impact that the pandemic has had on the relationship between PPAs and gateway communities.

Future research and stakeholder initiatives should focus on policy and community engagement strategy innovation. Much of the policy and philosophy that guides national park management in the United States was established when engagement with local communities was less of a priority, and, based on the results of this study and others, this can hinder effective collaboration. Hence, there is an opportunity to explore new policies which acknowledge and support the respective jurisdictions of the federal and local governments, while providing the flexibility needed for collaboration to flourish at the local level. Additionally, there is an opportunity to continue investigating collaborative initiatives, such as the ZLI initiative, to more accurately define effective engagement strategies and mechanisms that are universally applicable. These efforts will become increasingly important if resources available to support collaboration remain limited. Moreover, some of the collaborative initiatives described in this study are supported through less traditional mechanisms, such as corporate sponsorship and concessioner innovation. The implications of, and the degree to which, the Park Service and other PPA management agencies should rely on private support deserves more in-depth exploration.

Conclusion

This research provides insight into the outcomes and effects of PPA management approaches that are alternative relative to traditional PPA management strategies in the US NPS. As we shift from managing parks as islands to managing parks as part of larger environmental, socioeconomic and cultural landscapes, it will be important to understand how these changes are affecting quality of life and PPA-community relationships from the perspective of stakeholders experiencing these changes first hand. These results have management implications that may inform future community engagement and collaboration between the gateway community and managers at GTNP. Additionally, this case study provides insight into the need for a broader conversation on the current state and future of PPA policy regarding engagement and collaboration with gateway communities.

References

Acquah, E., Nsor, C.A., Arthur, E.K. and Boadi, S. (2017) The sociocultural impact of ecotourism on park-adjacent communities in Ghana. *African Journal of Hospitality, Tourism, & Leisure* 6(2), 1–14.

Andersen, P.H. and Kragh, H. (2010) Sense and sensibility: two approaches for using existing theory in theory-building qualitative research. *Industrial Marketing Management* 39(1), 49–55. DOI: 10.1016/j.indmarman.2009.02.008.

Bergstrom, R.D. (2018) Defining sustainability in the Greater Yellowstone ecosystem. *Journal of Sustainable Development* 11(1), 32–43. DOI: 10.5539/jsd.v11n1p32.

Bergstrom, R.D. and Harrington, L.M.B. (2018) Understanding agents of change in amenity gateways of the Greater Yellowstone region. *Community Development* 49(2), 145–160. DOI: 10.1080/15575330.2017.1416648.

Black, R. and Cobbinah, P.B. (2018) Local attitudes towards tourism and conservation in rural Botswana and Rwanda. *Journal of Ecotourism* 17(1), 79–105. DOI: 10.1080/14724049.2016.1258074.

Bricker, K.S. and Black, R. (2015) Sustainability in adventure programming: Maximizing benefits for cultural preservation, nature conservation, and local resident's well-being. In: Black, R. and Bricker, K.S. (eds) *Adventure Programming and Travel in the 21st Century*. Venture, Urbana, IL, USA, pp. 459–479.

Bruyere, B.L., Rodriguez, D.A. and Vaske, J.J. (2002) Enhancing importance-performance analysis through segmentation. *Journal of Travel & Tourism Marketing* 12(1), 81–95. DOI: 10.1300/J073v12n01_05.

Bushell, R. and Bricker, K. (2017) Tourism in protected areas: developing meaningful standards. *Tourism and Hospitality Research* 17(1), 106–120. DOI: 10.1177/1467358416636173.

Creswell, J.W. and Poth, C.N. (2016) *Qualitative Inquiry and Research Design: Choosing among five approaches*. Sage, Thousand Oaks, CA, USA.

Dinica, V. (2016) Protected areas: how will they contribute to third millennium challenges? *Policy Quarterly* 12(1), 50–59. DOI: 10.26686/pq.v12i1.4578.

Dunning, A.E. (2005) Impacts of transit in national parks and gateway communities. *Transportation Research Record* 1931(1), 129–136. DOI: 10.1177/0361198105193100116.

Eagles, P.F. and McCool, S.F. (2002) *Tourism in National Parks and Protected Areas: Planning and Management*. CAB International, Wallingford, UK.

Epler Wood, M. (2017) *Sustainable Tourism on a Finite Planet: Environmental, Business and Policy Solutions*. Routledge, London, UK.

Gabe, T. (2016) Effects of the October 2013 US Federal government shutdown on National Park gateway communities: the case of Acadia National Park and Bar Harbor, Maine. *Applied Economics Letters* 23(5), 313–317. DOI: 10.1080/13504851.2015.1071465.

Gandiwa, E., Heitkönig, I.M.A., Lokhorst, A.M., Prins, H.H.T. and Leeuwis, C. (2013) Illegal hunting and law enforcement during a period of economic decline in Zimbabwe: a case study of northern Gonarezhou National Park and adjacent areas. *Journal for Nature Conservation* 21(3), 133–142. DOI: 10.1016/j.jnc.2012.11.009.

Gregor, S. and Koithan-Louderbck, K. (1997) Marketing a resort community: Estes Park at a precipice. *Cornell Hotel and Restaurant Administration Quarterly* 38(6), 52–59.

Headwaters Economics (2018) *Montana's Economy, Public Lands, and Competitive Advantage*. Available at: https://headwaterseconomics.org/economic-development/trends-performance/montanas-economy-and-protected-lands/ (accessed 2 February 2022).

Keiter, R.B. (2013) *To Conserve Unimpaired: The Evolution of the National Park Idea*. Island Press, Washington, DC, USA.

Kerstetter, D.L. and Bricker, K.S. (2012) Relationship between carrying capacity of small island tourism destinations and quality-of-life. In: Uysal, M., Perdue, R., and Sirgy, M.J. (eds) *Handbook of Tourism and Quality-of-life Research: Enhancing the Lives of Tourists and Residents of Host Communities*. Springer, London, UK, pp. 445–462.

Lackey, N.Q. and Bricker, K.S. (2020) The relationship between Parks and protected areas and gateway communities in the United States and Canada: a scoping review. *Journal of Park and Recreation Administration*. DOI: 10.18666/JPRA-2020-10390.

Lindsey, P.A., Romañach, S.S., Matema, S., Matema, C., Mupamhadzi, I. *et al.* (2011) Dynamics and underlying causes of illegal bushmeat trade in Zimbabwe. *Oryx* 45(1), 84–95. DOI: 10.1017/S0030605310001274.

Marquit, J.D. and Mace, B.L. (2015) Park visitor and gateway community perceptions of mandatory shuttle buses. In: Orsi, F. (ed.) *Sustainable Transportation in Natural and Protected Area*. Routledge, London, UK, pp. 86–97.

Matose, F.M. (1994) Local people's uses and perceptions of forest resources: An analysis of a state property regime in Zimbabwe. Master's thesis. University of Michigan, Lansing, MI, USA.

Mbaiwa, J.E. and Stronza, A.L. (2010) The effects of tourism development on rural livelihoods in the Okavango Delta, Botswana. *Journal of Sustainable Tourism* 18(5), 635–656. DOI: 10.1080/09669581003653500.

Miles, M.B., Huberman, A.M. and Saldana, J. (2014) *Qualitative Data Analysis: A Methods Sourcebook*, 3rd edn. Sage, Thousand Oaks, CA, USA.

Morris, J.M. and McBeth, M.K. (2003) The New West in the context of extractive commodity theory: the case of bison-brucellosis in Yellowstone National Park. *The Social Science Journal* 40(2), 233–247. DOI: 10.1016/S0362-3319(03)00006-5.

Mose, I. and Weixlbaumer, N. (2007) A new paradigm for protected areas in Europe. In: Mose, I. (ed.) *Protected Areas and Regional Development in Europe: Towards a New Model for the 21st century*. Ashgate, Aldershot, UK, pp. 3–20.

National Park Service (1998) *The National Park Service Transportation Planning Guidebook*. Available at: https://www.nps.gov/orgs/1548/upload/transplan_Guidebook.pdf

National Park Service (2014) *Scaling Up: Collaborative Approaches to Large Landscape Conservation*. Available at: https://www.nps.gov/orgs/1412/upload/Scaling-Up-2014-508.pdf (accessed 2 February 2022).

National Park Service (2015) *A call to action:" Preparing for a Second Century of Stewardship and Engagement*. Available at: https://www.nps.gov/calltoaction/index.html (accessed 2 February 2022).

National Park Service (2019a) *Commercial Use Authorizations*. Available at: https://www.nps.gov/aboutus/commercial-use-authorizations.htm (accessed 2 February 2022).

National Park Service (2019b) *Learn About the Park*. GTNP, Wyoming. Available at: https://www.nps.gov/grte/index.htm (accessed 2 February 2022).

National Park Service (2020a) *Connected Conservation: What We Do*. Available at: https://www.nps.gov/subjects/connectedconservation/what-we-do.htm (accessed 2 February 2022).

National Park Service (2020b) *Visitation Numbers: 10 Most Visited Parks in 2019*. Available at: https://www.nps.gov/aboutus/visitation-numbers.htm (accessed 2 February 2022).

Puhakka, R., Cottrell, S.P. and Siikamäki, P. (2014) Sustainability perspectives on Oulanka National Park, Finland: mixed methods in tourism research. *Journal of Sustainable Tourism* 22(3), 480–505. DOI: 10.1080/09669582.2013.839690.

Ramos, A.M. and Prideaux, B. (2014) Indigenous ecotourism in the Mayan rainforest of Palenque: empowerment issues in sustainable development. *Journal of Sustainable Tourism* 22(3), 461–479. DOI: 10.1080/09669582.2013.828730.

Rose, J. and Johnson, C.W. (2020) Contextualizing reliability and validity in qualitative research: toward more rigorous and trustworthy qualitative social science in leisure research. *Journal of Leisure Research* 51(4), 432–451. DOI: 10.1080/00222216.2020.1722042.

Schelhas, J. (2009) The U.S. national parks in international perspective: the Yellowstone model or conservation syncretism. In: Polisciano, G. and Farine, O. (eds) *National Parks: Vegetation, Wildlife, and Threats*. NOVA Science, Hauppauge, NY, USA, pp. 83–103.

Schwandt, T.A. and Gates, E.F. (2017) Official science methodology is the study of how a particular kind of investigation should. In: Denzin, N.K. and Lincoln, Y.S. (eds) *The SAGE Handbook of Qualitative Research*. Sage, Thousand Oaks, CA, USA, pp. 341–358.

Slocum, S.L. (2017) Operationalising both sustainability and neo-liberalism in protected areas: implications from the USA's National Park Service's evolving experiences and challenges. *Journal of Sustainable Tourism* 25(12), 1848–1864. DOI: 10.1080/09669582.2016.1260574.

Snyman, S. and Bricker, K.S. (2019) Living on the edge: benefit-sharing from protected area tourism. *Journal of Sustainable Tourism* 27(6), 705–719. DOI: 10.1080/09669582.2019.1615496.

Stoker, P., Rumore, D., Romaniello, L. and Levine, Z. (2020) Planning and development challenges in Western gateway communities. *Journal of the American Planning Association* 87(3), 1–13. DOI: 10.1080/01944363.2020.1791728.

Tampakis, S., Karanikola, P., Tsantopoulos, G. and Andrea, V. (2014) The role of forest protected areas: a gateway community resident approach, case evidence from Cyprus. *Journal of Food, Agriculture & Environment* 4(1), 374–377.

Zajchowski, C., Lackey, N.Q. and McNay, G.D. (2019) "Now is Not the Time to Take a Breather": United States federal land management agency professionals' perceptions at the 40th anniversary of the Clean Air Act Amendments of 1977. *Society & Natural Resources* 32(9), 1003–1020. DOI: 10.1080/08941920.2019.1605436.

5 Advocacy for Protected Areas in the Caribbean: A Collaborated Approach to Ecotourism in Antigua and Barbuda

Therez B. Walker[1]* and Natalya Lawrence[2]
[1]University of the West Indies, Jamaica; [2]Offshore Islands Conservation Programme Coordinator, Environmental Awareness Group, Antigua and Barbuda

Abstract
The Caribbean is unique for its diverse marine and terrestrial environment that provides shoreline protection, food security and livelihoods, as well as facilitating tourism and other economic activities, for its residents and visitors alike. The role of Protected Areas (PAs) for conservation has been realized, and consideration must be given to the intricate linkages between social, economic and biological systems on small islands. The North East Marine Management Area (NEMMA) plays a significant role in the island of Antigua and Barbuda, for its importance to biodiversity conservation, as it connects many marine PAs in one continuous zone over the commercially important east coast of the island. This chapter highlights the advocacy work of those in gateway communities, and the importance of stakeholder collaboration in developing ecotourism. Practical implications emerge, as this case provides a framework for governance to collaborate for a more inclusive form of sustainable ecotourism in PAs.

Keywords: biodiversity, Caribbean, protected areas (PAs), livelihoods, ecotourism

5.1 Introduction

In 2019, travel and tourism contributed US$8.9 trillion to the world's gross domestic product (GDP), in addition to providing US$948 billion in capital investments, and 330 million jobs (World Travel and Tourism Council [WTTC], 2020a). Global tourism has been growing significantly over the last decade, and growth in the Caribbean region was recorded by the WTTC (2020b) at 3.4% in 2019. This is higher than the world average and illustrates the resilience of the world's most tourism-dependent region, as many Small Island Developing States (SIDS) were still recovering from devastating hurricanes that severely damaged tourism infrastructure in 2017. The steadfast economic performance of these small islands is inherently linked to the strength of their tourism sector (Scheyvens and Momsen, 2008; Seetanah, 2011), and the reality of being small also allows these countries to achieve flexibility by the ability to quickly adapt, innovate and respond to changes in market demands (Streeten, 1993; Croes, 2006; Scheyvens and Momsen, 2008). Tourism remains a mainstay for SIDS and accounts for a large share of their economies.

For the Caribbean, these small islands have had a long history of tourism dependency,

*Corresponding author: twalker00@gmail.com

 Tourism Transformations in Protected Area Gateway Communities (eds S.L. Slocum *et al.*)
DOI: 10.1079/9781789249033.0005

and their resilience has been tested by natural disasters and other crises situations (Alberts and Baldacchino, 2017). With the challenges brought on by the more recent Covid-19 pandemic, Caribbean SIDS have been severely impacted by reduced visitation. Nonetheless, because the region is heavily reliant on international tourism, the sector will play a major role in their economic recovery post Covid-19. With this, the awareness and sensitivity to nature and human connections may positively influence a more sustainable form of travel and tourism post pandemic (Galvani *et al.*, 2020).

The main foreign exchange earner for many Caribbean SIDS is tourism. In Antigua and Barbuda, the industry contributes 42.7% to GDP (US$ 1,432.5 million annually), and 90.7% of total employment (WTTC, 2020c). Globally, this ranks Antigua and Barbuda as seventh in terms of tourism contribution to GDP, and first in terms of tourism contribution to national employment. In 2019, Antigua and Barbuda recorded 300,990 stayover arrivals (Ministry of Finance and Corporate Governance, 2019), 734,000 cruise ship passengers and 20,600 yacht passengers (Antigua Hotel and Tourist Association, 2020; Eastern Caribbean Central Bank, 2020). The destination marketing organization efforts for Antigua and Barbuda proudly boast 365 beaches – a different beach for every day of the year. It is, therefore, fair to surmise that tourism in Antigua and Barbuda is reliant on its natural resources. Tourism activities largely include leisure at beaches, diving, snorkelling and other marine-based activities. However, when natural resources are overused as a result of tourism activities, sustainability becomes severely threatened, and as environmental conditions worsen, so does economic wellbeing, leading to an overall drop in visitor satisfaction (Briassoulis, 2002). Moreover, in recent years, escalating numbers of visitors and pleasure yachts have significantly increased human movement in Antigua and Barbuda's protected areas (PAs), which has increased the risk of invasive species being introduced in ecologically fragile areas on the islands (Daltry *et al.*, 2017). For these reasons, over the last few decades, marine PAs have been receiving considerable attention throughout the academic discourse.

The benefits and positive impacts of tourism to the gateway communities that neighbour PAs are many, and these include income from tour operations, in addition to other activities, such as commercial fishing, that supply goods to tourism establishments. It therefore becomes necessary for tourism to be managed in a way that offers a livelihood to those in isolated locations, and at the same time must balance with conservation efforts in a sustainable way. Progressively, PAs that have been established in recent years have focused on the role of local communities in environmental conservation (Nepal, 1997; Stevens, 1997). This is significant because these gateway communities act not only as beneficiaries of natural attractions, but also custodians, and it is essential to actively engage with these stakeholders for effective management of PAs (Strickland-Munroe, 2017).

In Antigua and Barbuda, the North East Marine Management Area (NEMMA) is a bustling hub of activity for subsistence fishers, along with local and overseas visitors eager to spend time in a pristine habitat, hoping for the opportunity to observe rare and unique aquatic species. Consistent research, habitat restoration and awareness-raising by volunteer groups have been the cornerstone of conservation efforts within the NEMMA (Jackson, 2008), and land-based conservation within the PA has had a positive impact on marine ecosystems (Environmental Awareness Group of Antigua and Barbuda, 2020).

The purpose of this chapter is to examine the relationship between stakeholders in PAs, as a way to foster collaboration for sustainable ecotourism in SIDS. This chapter is co-created with key stakeholders and uses action research with a mixed-method approach that combines residents' questionnaire responses and interviews. Taking stakeholder theory and collaboration into account, this chapter will analyse the views and advocacy work of stakeholders in gateway communities to explore a collaborative approach to PA management and provide an updated view on conservation for PAs in the Caribbean.

5.2 Literature Review

5.2.1 Tourism and sustainability in SIDs

The uniqueness of tourism in small islands has been discussed extensively throughout the tourism literature, inclusive of its opportunities and challenges (Briguglio, 1995; Weaver, 1995; Kerr, 2005; Scheyvens and Momsen, 2008). At the same time, the literature has not only exemplified how tourism allows small islands to overcome their size, but it also indicates how having a small population with tight kinship networks creates a strong sense of identity that encourages traditional approaches to development (Walker, 2017). Graci and Dodds (2010) have further shown how remoteness and traditional knowledge of adapting to changes help to strengthen communities. However, it is important to note that while tourism provides a means to overcome small size, tourism in island destinations does introduce various challenges. This has been examined by researchers, such as Butler (2015), who addresses the impact of human activity on natural environments, as well as contemporary issues relating to tourism and climate change. The significance of this to SIDs is great, as islands are extremely fragile and vulnerable tourism destinations that will continue to experience a variety of environmental, economic, and social pressures in the future (Carlsen and Butler, 2011).

It is in this vein that sustainable tourism development has been receiving increasing attention from policy makers in recent years; this has also been the case for the Caribbean. Walker (2020) thoughtfully discusses the evolution of sustainability in SIDs and Vallega (2007) clearly articulates that concerns for sustainable development in SIDs have been raised, specifically in ecological and social contexts. SIDs continue to face unique challenges relating to their insularity, as they are characterized by a fragile ecological makeup (Douglas, 2006). In particular, Caribbean small islands are highly dependent on the tourism industry (Walker, 2019), along with the coastal ecosystem for fisheries and shoreline protection, just to name a few (Lewsey *et al.*, 2004; Agostini *et al.*, 2015; Waite *et al.*, 2015). Still, Caribbean islands are facing accumulative challenges that relate to environmental degradation as a result of profit-making socioeconomic activities (Pantin, 1994), and this remains the case even though the academic discourse on SIDs has extensively discussed the dangers of this reality.

5.2.2 Tourism in marine PAs – the case for ecotourism

One common definition of a marine PA is "any area of intertidal or subtidal terrain, together with its overlying water and associated flora, fauna, historical and cultural features, which have been reserved by law or other effective means to protect part or all of the enclosed environment" (Kelleher and Kenchington, 1991, p. 7). While the aim of protected coastal areas is to preserve ecosystems, conservation is coupled with a range of socioeconomic activities, such as tourism (de Fontaubert *et al.*, 1996). Here, the need for management of these multiple-use areas becomes apparent, and these marine PAs are commonly referred to as marine managed areas. The benefits derived from management and conservation of marine PAs ultimately leads to desirable impacts for gateway communities, as they reap a variety of benefits in the fisheries sector (Gell and Roberts, 2003), as well as awareness of negative ecological impacts in the Caribbean (Wise, 2014). Additional benefits to the tourism industry and sustainable livelihoods of residents are experienced, although researchers have noted that marine PAs have been criticized for overuse leading to negative impacts (Bennett and Dearden, 2014).

It, therefore, becomes necessary for stakeholders to deliberate on the ultimate goal of tourism development, while also embarking towards sustainability (Wondirad *et al.*, 2020). For small islands, it is important to remember that development is multifaceted, and the socioeconomic aspect of sustainable development should be included in the overall process (Walker and Lee, 2019). Since sustainability is a matter of both local and global responsibilities, it is important to realise that it is not a one-dimensional phenomenon in the global nexus (Saarinen, 2006). Furthermore, by including all aspects of sustainable development into the tourism discourse for SIDs, it provides an opportunity to formulate better guidelines for making gateway communities more resilient, in the face of various challenges (Scott *et al.*, 2012; Timms and Conway, 2012; Walker, 2017).

Stronza and Gordillo (2008) have expressed that, although tourism is notoriously known for its potential to disrupt natural habitats and local communities, ecotourism can be an incentive for conservation, particularly when it ignites positive economic benefits. Therefore, ecotourism is often seen as an alternative type of tourism development that has the potential to achieve the dual aims of conservation and sustainable development (Wondirad *et al.*, 2020). Furthermore, Caribbean small islands have recognized marine PAs as valuable tools, for not only conservation, but also for the recovery of key habitats in decline, since a long history of overuse for commercial purposes and poor coastal development have had a negative impact in many coastal areas throughout the region (Bustamante *et al.*, 2014). Ecotourism could make a valuable contribution to conservation and biodiversity by providing an economic return on activities that can change unsustainable land use practices into low impact activities (McKercher, 2010).

5.3 Case Study – North East Marine Management Area (NEMMA)

NEMMA was declared a marine reserve in 2005, comprises more than 30 coralline islands, and encompasses an area of over 30 square miles in the Atlantic Ocean. These islands represent several threatened ecoregions that include the Leeward Islands dry forests, xeric shrub and mangroves, which are severely threatened due to coastal hotel developments and pollution in Antigua and Barbuda. Interestingly, the NEMMA is also considered to be a globally significant research and conservation site for land animals within this key biodiversity area, which include most of the country's endemic reptiles, such as the world's population of the critical Antiguan Racer snake (*Alsophis antiguae*) (Jackson, 2008). Further, with the absence of light pollution, and efforts to remove invasive alien species from some offshore islands, their beaches provide nesting habitat for critically endangered Hawksbill turtles (*Eretmochelys imbricata*) and vulnerable Leatherback turtles (*Dermochelys coriacea*), in addition to endangered Green turtles (*Chelonia mydas*). Flora include a yellow *Talinum cf. fruticosum* (found only on the offshore islands), endangered *Lignum vitae*, and many other species listed as nationally threatened (Pratt *et al.*, 2009).

5.3.1 Advocacy for conservation and sustainable tourism

Conservation work began on offshore islands situated in the NEMMA before it became a PA. This advocacy work began in 1995, following the rediscovery of the Antiguan Racer snake, and a bid to save it. Grassroots non-governmental organizations (NGOs), community members, local government partners, as well as regional and international NGOs joined forces to save the critically endangered snake, endemic only to Antigua and Barbuda. Given the outstanding success of these advocacy efforts, not only did the snake population expand, a whole host of plants and other animals benefitted as well. Historical data from the NEMMA reveal restoration, where the offshore islands experience a bustling, thriving rejuvenation, and data from the last 25 years further disclosed that visitation to the islands increased exponentially within that time frame. Furthermore, the offshore islands are areas of outstanding beauty, unmatched vistas, and what is often referred to as the last local refuge for endemic wildlife (Espeut, 2006; Environmental Awareness Group of Antigua and Barbuda, 2020). It is often included in the top 10 sites to visit in Antigua and Barbuda, and a must visit site for ecotourism enthusiasts.

Stakeholders with a vested interest in the NEMMA urged for the key biodiversity area to be designated a PA, and in 2005 it was officially designated as the largest PA in Antigua and Barbuda at 3,600 hectares. From the NEMMA Management Plan (Jackson, 2008), it is clear that the area's governing authority is the government. However, it also makes clear that civil society and other stakeholders must share in the responsibility to manage the PA. The NGO, Environmental Awareness Group of Antigua and Barbuda (EAG) has been the de facto terrestrial-based manager, collaborating with tour operators, resident, and several government departments. The group, with its strong network of partners, volunteers, and community supporters, has executed several successful conservation projects within the

NEMMA. Through the conservation efforts of the EAG, citizen scientists have gained expertise to monitor the PA and training in areas such as invasive alien species detection and control. Biodiversity monitoring (including reptiles and avifauna) has assisted residents in understanding trends in conservation management that guide conservation decision making. Here, the NEMMA provides educational awareness and institutional learning for its residents and is considered a living lab suitable for ground-breaking research.

The main threat to this PA is alien invasive species - most notably Eurasian black rats (*Rattus rattus*) which, by the mid-1990s, occupied many islands in this ecologically fragile area. It is important to note that visitors arriving by boat to the offshore islands are a primary pathway for the introduction of this alien species that has no natural predators in the isolated ecosystem. Thus, the priority of stakeholders became the eradication of the alien species, in order to save its prey, the Antiguan Racer, from extinction. In addition to successfully saving this rare reptile, other unanticipated benefits were derived: increase in nesting bird pairs of both resident and neotropical migrants; and the resurgence of rare, and in some cases assumed extinct, plants (golden *Tallinum fructicosum* sp.). The vegetation biomass increased across the islands, such as Great Bird Island, on removing the invasive alien species *Rattus rattus*, and in some cases *Herpestes javanicus*, to safeguard the survival of the racers. Further, increases in other reptiles, such as three species of endemic lizards (*Anolis wattsi*, *Anolis leachii*, *Pholidoscelis griswoldi*) (Ross, 2010, 2011; Coldwell, 2017) and an increase in turtle nesting activity, were observed.

Ultimately, alien species eradication has had astonishing consequences, and community volunteers have made observations, along with data collection, which have indicated that native plant biomass has doubled, native terrestrial reptile biomass has increased threefold, nesting Hawksbill turtles have increased fourfold, and particular bird species have increased from three up to twentyfold. The Antiguan Racer increased by twenty-fourfold to more than 1200 individuals, all on islands within the NEMMA that have been cleared of the invasive species. This is evidenced by regular data collection in the form of fixed-point photos on various points on several islands. At the same time, from a tourism perspective, the number of people using and enjoying the restored islands has risen from 17,000 to over 50,000 per year, bringing more than US$5 million to the local economy from pleasure yachts alone.

Antigua and Barbuda represent an important case study for a combination of reasons. Located in one of the most tourism dependent regions in the world, this Caribbean island has focused on sustainable tourism development, particularly from an environmental perspective. Additionally, the efforts of volunteer groups and those in gateway communities, with their consistent mandate towards environmental law and policy (EAG, 2019), have been successful in advocating for recent legislation, as they are positioned to challenge the rush to develop tourism projects around fragile wetlands and rare species. Bramwell and Lane (2011) emphasize that governance cannot be understood in isolation from its relationship with people, including community groups that seek to influence the governance process. This highlights the importance of how advocacy groups can effect change in small islands, and in Antigua and Barbuda, many small influential groups neighbouring the PA have contributed significantly in promoting environmental and sustainability issues, although they generally mobilize to address specific issues relating to their surrounding community areas.

5.4 Data Collection

Data collection for this study consisted of a document study, as well as primary data collection through questionnaires and interviews. The document study and careful review of literature provided the theoretical groundwork of this research in regards to the role of advocacy and stakeholder collaboration in PAs. The researchers utilized an electronically distributed survey questionnaire, and in-depth stakeholder interviews were conducted to gather further insights from stakeholders in gateway communities that neighbour the marine PA. Residents who border the PA returned 76 completed questionnaires, and the questionnaires provide several key insights. These include demographic information, awareness level of the PA by community members, the perception of impacts on biodiversity and livelihoods in the PA, and the attitudes of those in the gateway community towards the development of a special economic zone (SEZ) within the PA.

To gain additional information, the researchers conducted additional in-depth interviews with 16 stakeholders within the gateway community. The targeted stakeholder groups for the interviews were residents who lived in the area for more than ten years, commercial fisherfolk, leisure and recreational stakeholders, as well as small- and medium-sized business owners. The interviews focused on documenting the socioeconomic impacts of those who reside within the PA and depend on the PA for their livelihood. In addition to the socioeconomic importance of tourism to the area, the interviewees discussed issues relating to advocacy, as well as stakeholder collaboration that forms the motive for ecotourism in the PA. Each stakeholder was selected based on their expertise in relation to their engagement with the development of sustainable tourism in the PA. However, to protect the study from being overly expansive, and at the same time ensure its validity, only four stakeholders from each group were interviewed. The main objectives of this study were to:

- explore the importance of sustainable tourism for gateway communities in SIDs;
- analyse the impacts of PAs on communities' lives and livelihoods; and
- examine the implications of advocacy and stakeholder collaboration for achieving sustainability in PAs.

5.5 Data and Analysis

Researchers have identified stakeholder groups to include employees, customers, suppliers, lenders, shareholders, government and non-governmental organizations, and residents (Sheehan and Ritchie, 2005). Figure 5.1 recognizes that for efficient implementation of stakeholder management in PAs, there must be a genuine appreciation of all groups or individuals with an interest in the multiple-use area. We have provided some modifications to stakeholders customarily identified in the tourism

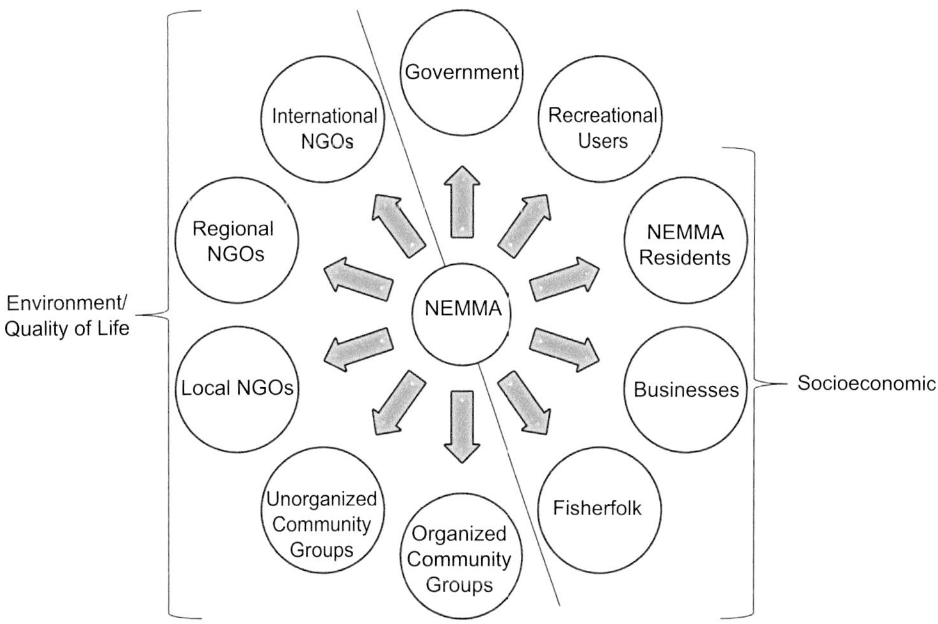

Fig. 5.1. Ecotourism stakeholders map in the NEMMA (figure courtsey of the authors).

literature to reveal the nature of the stakeholder involvement for dynamic tourism that takes place in PAs, such as NEMMA. Specifically, our graphical representation includes those with a primary socioeconomic interest in NEMMA, such as its residents, businesses, and fisherfolk located in gateway communities. It is important to mention that fisherfolk and small business owners (e.g. restauranteurs, tour operators) are also NEMMA residents. This also applies to those with a mainly environmental interest in the NEMMA, since advocacy groups are often made up of residents as well.

This investigation revealed that NEMMA residents are quite knowledgeable about biodiversity within the PA, the importance of the PA, as well as how conservation positively impacts their livelihoods. Seventy-nine per cent of residents were aware of the Antigua Racer and could identify the West Indian whistling duck, in addition to other marine life found in the NEMMA. This is due, in part, to the educational aspect of the NEMMA, whereby residents are trained as ecotour guides.

Residents have a vested interest in the environment, and when asked about how conservation efforts in the PA impact local livelihoods, the responses were varied. Residents were mainly concerned about mangrove destruction and subsequent issues of shoreline protection. This may, in part, be due to the recent hurricanes that have affected the Caribbean, yet at the same time, residents also indicated that destruction of the ecosystem could negatively impact NEMMA's efforts to provide a breeding ground for fish. As much as 70% of respondents felt that the biodiversity and wildlife in the NEMMA was instrumental in positively impacting their livelihoods, and the inherent value placed on the environment by the gateway community may explain similar sentiments shared about advocacy by the community towards conservation.

Additional in-depth interviews found that, while employment within the hospitality sector was commonplace for many NEMMA residents, many individuals appreciate residing in close proximity to the marine environment, as it provides alternative sources of income. Here, they engage in recreational and commercial fishing, and boat owners engage in tour operations due to their familiarity of the area. The researchers were able to gain further insights into the socioeconomic impacts of the NEMMA on key stakeholders and how the lives and livelihoods in the community are dependent on the tourism industry, in particular ecotourism, due to the unique nature of the NEMMA as a marine PA. Interviewees referred to the diverse economic opportunities that were available to many in the gateway community due to its coastal location and the thriving tourism industry in Antigua and Barbuda. These opportunities ranged from water tour guide, boat operator, fishing, dive operator, restaurateur/cafe, Airbnb operator, taxi guide, among others, that provide residents of the community an alternative to low wages in neighbouring hotels.

Through semi-structured interviews, stakeholders in the gateway community also shared their views on the role of advocacy, their opportunities and challenges relating to stakeholder engagement, as well as their outlook on ecotourism as a potential solution. Key findings reveal that stakeholders view advocacy as a traditional approach to community mobilization. For residents of the gateway community, advocacy is twofold: it is a response to protecting their quality-of-life values where they utilize the area for recreational/leisure purposes; and, just as equally, it is a response to protecting socioeconomic values where their livelihoods are dependent on the ecosystem of PA. This study of the NEMMA stakeholders also revealed that, while some active community members are aware of the PA, with only 34% knowing its specific geographical location, there is a need for more awareness. Enforcement was also cited as a challenge being faced since common violations of overfishing, mangrove destruction, or infrastructural development that harm the ecosystem, often go unchecked by the relevant authorities.

Ecotourism in the NEMMA has been driven by advocacy, and it is the form of tourism that is currently being practised as a model that can be replicated in other parts of Antigua and Barbuda. Stakeholders in the NEMMA view ecotourism as a model that fits the unique needs of the gateway community and as a potential solution to issues they face. Notwithstanding, this investigation has revealed that there is an overwhelming concern that recent coastal development is destroying the progress made in developing the ecotourism product.

It is important to note that SIDs, such as Antigua and Barbuda, are traditionally known as 3S (sun, sea, sand) destinations. Given that the NEMMA is a marine PA, tourism activity is quite significant in the area, with resorts and other infrastructure developments along the coast. NEMMA has been the site of several development projects that have been met with public controversy owing to destruction of natural habitats (Jackson, 2008). Thereby, the advocacy work of residents and other stakeholders have prevented several developments in the area to date. Most recently, development of a SEZ in the NEMMA has received attention from advocacy groups who had successfully lobbied for legislation that protects the NEMMA from development that negatively impact the PA. Nonetheless, with such high economic value to the government, subsequent legislation now overrides the functioning of the Environmental Protection and Management Act (2015) and allows for a SEZ. Advocacy groups have highlighted aggressive destruction of mangroves in the NEMMA to facilitate this development, however enforcement of the remaining applicable laws has been weak. Although 37% of the residents surveyed indicated that they do not know how they feel about this SEZ, 44% saw this as bad for the PA.

Fig. 5.2. Advocacy for stakes in the NEMMA (figure courtesy of the authors).

5.6 Discussion

In the NEMMA, the collaborated efforts of stakeholders from a grassroots level have allowed for capacity building and supportive networks that positively impact coastal tourism. Accordingly, stakeholder theory is recognized as an important consideration for this analysis due to the dynamic nature of tourism in SIDs, and it lends support to a more practical application of the sustainable development concept. Stakeholder theory can be traced back to 1759 (Mainardes *et al.*, 2011), although it was detailed by Freeman (1984), when he addressed the concept of globalization, by proposing that an organization is characterized by its interconnecting relationships within diverse groups. Various stakeholder concepts, such as corporate planning, systems theory, corporate social responsibility, and organizational theory all play a major role in the formulation of stakeholder theory (Freeman and Mc Vea, 2001), yet, one of the main elements argued is that decision making should consider the interests of all stakeholders.

Figure 5.2 illustrates the main interests of NEMMA stakeholders. Stakeholders are grouped according to their priority values, such as socioeconomic values, where livelihoods, self-sufficiency, skills enhancement, social progress, and appreciation are achieved. Similarly, quality of life priority values materializes through advocacy efforts, such as awareness, community pride, engagement, wellbeing, capacity building and self-identity. The government and recreational stakeholder groups have a stake in each value and the three E's of sustainable tourism (economy, environment, equity).

The current social economy and ecotourism activities require a bottom-up approach, and there has been strong environmental advocacy in the NEMMA by NGOs, other pressure groups, and residents, who also have a socioeconomic interest in the PA. However, even though residents represent a large portion of the stakeholders, and are custodians of the PA, they have unfortunately been excluded from the decision-making process for the development of hotels and other investments that have a high economic value to the government (Espeut, 2006; Jackson, 2008; OECS, 2008). Sautter and Leisen (1999) discuss how consideration should not only be given to all stakeholder groups, but relative power should also be neutralized. There appears to be balance of power issues unfolding in the NEMMA. If these issues are defused, stakeholders with quality of life interests will be better able to voice concerns and collaborate for low impact ecotourism in the NEMMA.

The tourism industry is also one where a diverse range of stakeholders need to collaborate, communicate and build partnerships even though they may have conflicting interests. Indeed, the tourism discourse has acknowledged a diverse range of stakeholder groups in this dynamic industry (Robson and Robson, 1996; Sautter and Leisen, 1999; Hardy, 2005; Sheehan and Ritchie, 2005; Timur and Getz, 2008; d'Angella and Go, 2009; Tkaczynski *et al.*, 2009). At the same time, when tourism in PAs is considered, it is important to balance tourism activity with conflicting interests of stakeholders and the natural resources needed to sustain the industry. It is for this reason that, within the context of sustainable tourism development, it becomes central to first identify the key stakeholders with a vested interest, before it is possible to examine how they influence the decision-making process involved in managing ecotourism in a PA.

Stakeholder theory, from a tourism perspective, suggests that policy makers integrate stakeholder considerations into the planning process. Considering the stakeholder management process according to Freeman (1984), the key to achieving this is following four steps, and when applied to tourism with a sustainable tourism emphasis, Theobald (2005) suggests that the four steps include:

- identify relevant stakeholder groups in relation to sustainable tourism development;
- determine the stakes (potential interest) of stakeholders;
- determine how effectively stakeholder expectations are currently being met; and
- modify the tourism development policies and priorities to take into consideration stakeholder interest that are not currently met.

Since stakeholder groups have varying interests, this creates a challenge in the planning and management of ecotourism in PAs. It is, therefore, important for all stakeholder groups to align their objectives and take a more cohesive approach to gaining value. Here, stakeholder collaboration becomes an essential element to resolving any issues, and obtaining, compromising and building trust amongst key players (Walker, 2017). Furthermore, a stakeholder theory of collaboration in PAs should integrate the relationship between all organizations, the natural area, those who inhabit it, and those who have a stake in it (Jamal and Stronza, 2009). The common interest in the NEMMA, and the advocacy surrounding it, serves as the catalyst bringing stakeholders together.

Socioeconomic values and quality of life values are realised through the advocacy work of stakeholders. Yet, while considering the inclusive mandate of the stakeholder management process for the PA, this investigation of the NEMMA agrees with previous studies (Jackson, 2008; OECS, 2008) in determining how stakeholders' expectations are being met. Few studies have identified stakeholder concerns as they relate to the management of the NEMMA. Secondary data, questionnaire responses, in addition to in-depth interviews, exposed that stakeholders did not feel included in the decision-making process, and advocacy groups are outspoken about their disproval of contemporary developments in the gateway community. There has also been a lack of enforcement for existing legislation, as well as lack of authority to report/penalize persons who engage in unregulated activities in the PA. Figure 5.3 illustrates a framework of stakeholder involvement for ecotourism in the NEMMA.

Notwithstanding, while much of the tourism/recreational uses in the NEMMA are centred around restaurants, beach recreation, swimming, ecotours, diving, snorkelling, yachting, kayaking, vending of craft/souvenirs, and other low impact water sports, these are all dependent on the overall health of the ecosystem in the PA. Thus, conservation of all aspects of the fragile marine and terrestrial environment is key for continued multiple-use, inclusive and low impact ecotourism.

5.7 Practical Implications

This study has sought to provide a more meaningful research approach to include the views of persons in gateway communities within the SIDs network. The interview and survey of a variety of stakeholders within the NEMMA

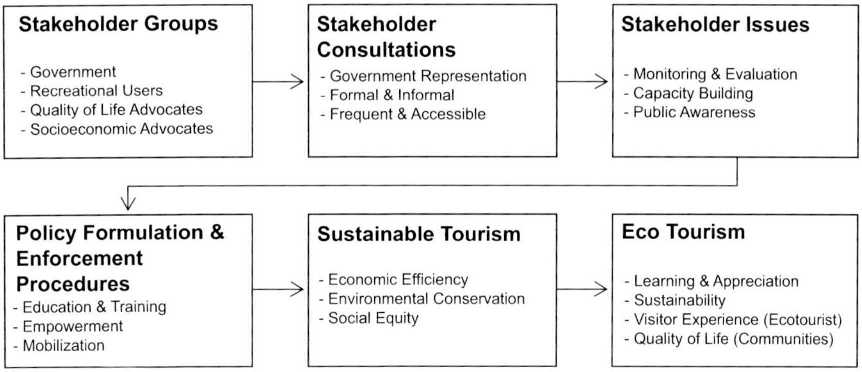

Fig. 5.3. Framework of stakeholder involvement for ecotourism in the NEMMA (figure courtesy of the authors).

found that respondents were receptive to the PA and recognized its importance. Although these stakeholders in the gateway community are aware of the importance of the PA, this study shows that while capacity building was commonplace, more awareness, access to participation and enforcement of existing regulations within the PA are needed. It is, therefore, recommended that emphasis be placed on these areas going forward, as this will allow for additional ownership by those in the gateway community who act as guardians of the PA. Additionally, there is room to learn from the case of gateway communities in the NEMMA, as the Caribbean and other SIDS progressively move towards a more sustainable tourism product. Here, policy makers will need to reflect on the overarching framework that will be necessary for regional collaboration on sustainability.

Furthermore, with the general push towards sustainable development in SIDS, alternative forms of tourism, such as ecotourism, becomes practicable. With multiple aims, such as engaging communities, conserving the environment and providing productive jobs, this chapter seeks to highlight theoretical and practical implications for low impact ecotourism in PAs. The traditional approach to advocacy has encouraged custodians of the environment, protecting not only their livelihoods, but also utilizing their collective knowledge of biodiversity that ultimately assist in minimizing the negative impacts and maximizing the positives. Capacity building initiatives that have been implemented by pressure groups, such as the EAG, ensures that stakeholders are not only mobilized through advocacy, but that they gain skills and participate in data collection, monitoring, evaluation and learning within the NEMMA. Previous tourism studies have revealed that residents in small Caribbean coastal communities have mobilized pressure groups to seek alternative approaches to development, and these communities have been successful in initiating bottom-up approaches that allow them to continually adapt for improvement in their quality of life, while fully participating in the development process (Walker, 2019). However, the examples of community-led initiatives in Caribbean small islands emphasize a different reality from the areas that are supported by policy makers. In fact, although policy documents frequently highlight the importance of people in the development process, there is very little support that is given to the process of implementation (Pantin, 1994; Apostolopoulos and Gayle, 2002; Singh, 2008). This is evident by the contemporary issues being faced with the development of a SEZ in the PA, which comes after decades of advocacy work by stakeholders. In 2015, the introduction of an environmental law, which was a policy intended to specifically protect the NEMMA from developing tourism projects around the fragile ecosystem, has all but been ignored.

Sustainability is essentially integrated within the characteristics of SIDs, and for PAs, where the focus is on environmental conservation, stakeholders can collaborate to provide the learning necessary to achieve ecotourism. The capacity building initiatives, coupled with the value placed on the connection between the environment and the livelihoods of residents, have allowed for some knowledge of biodiversity and ecosystem importance. Scholars, such as Armitage *et al.* (2008), have discussed the challenges with learning in complex systems involving adaptive and collaborative management efforts, since the extent to which these efforts shape learning greatly depend on the strategies employed. In the NEMMA, multiple stakeholders, such as NGOs, residents, fisherfolk and others, have participated in capacity building initiatives where collaborated management is supported by communication and knowledge sharing. The ultimate goal of adaptive co-management is to ensure that all actors ultimately perceive the system as integrated, complex and adaptive through dialogue and feedback between resource users on resources, ecosystems, governance and public infrastructure (Fabricius and Currie, 2015). Yet, as conflict arises, new forms of governance should emerge (Jamal and Stronza, 2009). Therefore, the aim going forward should focus on creating arenas for addressing current issues relating to enforcement of violations in the NEMMA. This, in combination with stakeholder consultations, should result in trust building, knowledge generation, collaborative learning, preference formation and conflict resolution amongst stakeholders as it relates to environmental issues (Odon Green *et al.*, 2015) and alternative forms of tourism, such as low impact ecotourism activities in PAs.

5.8 Conclusion

Gateway communities are important for a number of reasons, primarily they provide scenic and authentic spaces that are ideal for small island tourism. Yet, it is equally important that quality of life values are balanced with socioeconomic values so that there is a coexistence of humans and nature. The case study in this chapter has highlighted that advocacy for the fragile marine PA has resulted in low impact ecotourism. Advocacy here has resulted in mobilization, improved knowledge and environmental monitoring by citizen scientists (residents) along with other stakeholders. This coupled with the mutual benefits derived by various groups has resulted in continued dialogue between these stakeholder groups. Moreover, since the NEMMA, as a marine space, includes several pristine offshore islands, it has reserved its spot as the number one tourist attraction in the country for ecotourists.

This has allowed stakeholders to benefit from the PA, while simultaneously monitoring impacts and providing educational tours that increase awareness of the cherished landscapes that surround the gateway communities. Surveys and interviews with stakeholders in the gateway communities have highlighted several features of the NEMMA that are of importance to the community, and which have shaped contemporary issues in the gateway communities bordering the NEMMA. While there are residents who utilize the natural resources available for commercial fishing, tours and other activities, many residents do not, and this offers an opportunity for increased capacity building, as identified in this study.

This study has also provided a clear indication that the gateway communities in Antigua and Barbuda have already implemented successful initiatives that focus on safeguarding the natural environment while embracing low impact activities. Indeed, the literature has highlighted that when gateway communities advocate to protect their quality of life, socioeconomic wellbeing is also protected. Therefore, this case provides a learning opportunity for other gateway communities throughout the SIDs network, as it offers a practical approach on how these communities can collaborate and mobilize in ways that encourage inclusive and sustainable low impact tourism development. As groups engage for mutual benefit, stakeholder consultations identify key issues that create the basis for policy formulation. With this approach, unique needs are addressed and supported with training, monitoring and learning. Successful advocacy and resulting initiatives in the NEMMA help to motivate and facilitate a developmental approach that utilizes various stakeholders for the advancement of livelihoods, and the gateway communities themselves are able to formulate their unique approach towards ecotourism in the PA. With an already vibrant tourism industry, those with a stake in the NEMMA can use the opportunity to further the sustainability agenda and increasingly contribute to their livelihoods, making their community more resilient and sustainable.

References

Agostini, V.N., Margles, S.W., Knowles, J.K., Schill, S.R., Bovino, R.J. *et al.* (2015) Marine zoning in St. Kitts and Nevis: a design for sustainable management in the Caribbean. *Ocean & Coastal Management* 104, 1–10.

Alberts, A. and Baldacchino, G. (2017) Resilience and tourism in islands: insights from the Caribbean. In: Butler, R.W. (ed.) *Tourism and Resilience*. CAB International, Wallingford, UK, pp. 150–162.

Antigua Hotel and Tourist Association (2020) *Tourism statistics for Antigua and Barbuda*. Available at: http://members.antiguahotels.org/tourism-statistics-for-antigua-and-barbuda/ (accessed 3 February 2022).

Apostolopoulos, Y. and Gayle, D.J. (eds) (2002) *Island Tourism and Sustainable Development: Caribbean, Pacific, and Mediterranean Experiences*. Greenwood Publishing Group, Westport, CT, USA.

Armitage, D., Marschke, M. and Plummer, R. (2008) Adaptive co-management and the paradox of learning. *Global Environmental Change* 18(2008), 86–98.

Bennett, N.J. and Dearden, P. (2014) Why local people do not support conservation: community perceptions of marine PA livelihood impacts, governance and management in Thailand. *Marine Policy* 44(2014), 107–116.

Bramwell, B. and Lane, B. (2011) Critical research on the governance of tourism and sustainability. *Journal of Sustainability* 19(4-5), 411–421.

Briassoulis, H. (2002) Sustainable tourism and the question of the commons. *Annals of Tourism Research* 29(4), 1065–1085.

Briguglio, L. (1995) Small island developing states and their economic vulnerabilities. *World Development* 23(9), 1615–1632.

Bustamante, G., Canals, P., di Carlo, G., Gomei, M., Romani, M. *et al.* (2014) Marine PAs management in the Caribbean and Mediterranean seas: making them more than paper parks. *Aquatic Conservation: Marine and Freshwater Ecosystems* 24(2), 153–165.

Butler, R. (2015) The evolution of tourism and tourism research. *Tourism Recreation Research* 40(1), 16–27.

Carlsen, J. and Butler, R. (eds) (2011) *Island Tourism: Towards a Sustainable Perspective*. CAB International, Wallingford, UK.

Coldwell, V. (2017) Report on Seabird, Landbird and Lizard Surveys for Antigua's Offshore Islands, May – June 2017. Report to the Environmental Awareness Group, Antigua.

Croes, R.R. (2006) A paradigm shift to a new strategy for small island economies: embracing demand side economics for value enhancement and longterm economic stability. *Tourism Management* 27(3), 453–465.

Daltry, J.C., Lindsay, K., Lawrence, S.N., Morton, M.N., Otto, A. *et al.* (2017) Successful reintroduction of the critically endangered Antiguan racer *Alsophis antiguae* to offshore islands in Antigua, West Indies. *International Zoo Yearbook* 51, 1–10.

de Fontaubert, A.C., Downes, D.R. and Agardy, T.S. (1996) Biodiversity in the seas: implementing the convention on biological diversity in marine and coastal habitats. IUCN Environmental Policy and Law Paper No. 32. A marine conservation and development report. IUCN, Gland, Switzerland.

Douglas, C.H. (2006) Small island states and territories: sustainable development issues and strategies-Challenges for changing islands in a changing world. *Sustainable Development* 14, 75–80.

d'Angella, F. and Go, F.M. (2009) Tale of two cities' collaborative tourism marketing: Towards a theory of destination stakeholder assessment. *Tourism Management* 30, 429–440.

Eastern Caribbean Central Bank (2020) *Statistics. Tourism data*. Available at: http://www.eccb-centralbank.org/Statistics/#tourismdata (accessed 3 February 2022).

Environmental Awareness Group of Antigua and Barbuda [EAG] (2019) *Law & Policy*. Available at: http://www.eagantigua.org/page89.html (accessed 3 February 2022).

Environmental Awareness Group of Antigua and Barbuda [EAG] (2020) *Offshore Islands Conservation Programme*. Available at: https://www.eagantigua.org/page557.html (accessed 3 February 2022).

Environmental Protection and Management Act (2015) *No. 11 of 2015*. Available at: http://laws.gov.ag/wp-content/uploads/2019/02/a2015-11.pdf

Espeut, P. (2006) *Opportunities for Sustainable Livelihoods on One PA in each of the Six Independent OECS Territories, for the OECS PAs and Sustainable Livelihoods (OPAAL) Project*. Available at: http://parkscaribbean.net/wp-content/uploads/2013/11/Opportunities%20for%20Sustainable%2

0Livelihoods%20in%20Protected%20Areas%20in%20the%20OECS%20(2006).pdf (accessed 3 February 2022).
Fabricius, C. and Currie, B. (2015) Adaptive Co-Management. In: Allen, C.R. and Garmestani, A.S. (eds) *Adaptive Management of Social-Ecological Systems*. Springer, London, UK, pp. 147–179.
Freeman, R.E. (1984) *Strategic Management: A Stakeholder Approach*. Pitman, Boston, MA, USA.
Freeman, R.E. and Mc Vea, J. (2001) A Stakeholder Approach to Strategic Management. Darton Business School Working Paper No. 01-02.
Galvani, A., Lew, A.A. and Perez, M.S. (2020) COVID-19 is expanding global consciousness and the sustainability of travel and tourism. *Tourism Geographies* 22(3), 567–576. DOI: 10.1080/14616688.2020.1760924.
Gell, F.R. and Roberts, C.M. (2003) Benefits beyond boundaries: the fishery effects of marine reserves. *Trends in Ecology & Evolution* 18(9), 448–455.
Graci, S. and Dodds, R. (2010) *Sustainable Tourism in Island Destinations*. Earthscan, London, UK.
Hardy, A. (2005) Using grounded theory to explore stakeholder perceptions of tourism. *Journal of Tourism and Cultural Change* 3(2), 108–133.
Jackson, I. (2008) Northeast Marine Management Area (NEMMA) 2007-2010 Management Plan. Organisation of Eastern Caribbean States (OECS) Environment and Sustainable Development Unit (ESDU).
Jamal, T. and Stronza, A. (2009) Collaboration theory and tourism practice in PAs: stakeholders, structuring and sustainability. *Journal of Sustainable Tourism* 17(2), 169–189.
Kelleher, G. and Kenchington, R. (1991) *Guidelines for Establishing Marine PAs. A Marine Conservation and Development Report*. IUCN, Gland, Switzerland.
Kerr, S.A. (2005) What is small island sustainable development about. *Ocean & Coastal Management* 48, 503–524.
Lewsey, C., Cid, G. and Kruse, E. (2004) Accessing climate change impacts on coastal infrastructure in the Eastern Caribbean. *Marine Policy* 28(2004), 393–409.
Mainardes, E.W., Alves, H. and Raposo, M. (2011) Stakeholder theory: Issues to resolve. *Management Decision* 49(2), 226–252.
McKercher, B. (2010) Academia and the evolution of ecotourism. *Tourism Recreation Research* 35(1), 15–26.
Ministry of Finance and Corporate Governance (2019) *Tourism Statistics Bulletin*. Antigua and Barbuda Statistics Division. Available at: https://statistics.gov.ag/subjects/travel-and-tourism/ (accessed 3 February 2022).
Nepal, S.K. (1997) Sustainable tourism in PAs and livelihood need of local communities in developing countries. *International Journal of Sustainable Development and World Ecology* 4(2), 123–135.
Odon Green, O., Schultz, L., Nekoro, M. and Garmestani, A.S. (2015) The role of bridging organizations in enhancing ecosystem services and facilitating adaptive management of social-ecological systems. In: Allen, C.R. and Garmestani, A.S. (eds) *Adaptive Management of Social-Ecological Systems*. Springer, London, UK, pp. 107–122.
Organisation of Eastern Caribbean States [OECS] (2008) *Environmental and Socio-economic Studies for Opaal Demonstration Sites: Northeast Marine Management Area (NEMMA) Antigua.*Site report. ECO Report NO. 10/2007. Available at: https://parkscaribbean.net/wp-content/uploads/2013/11/North-East%20Marine%20Management%20Area-Environmental%20and%20Socio-economic%20Study%20(2007).pdf (accessed 3 February 2022).
Pantin, D. (1994) *The Economics of Sustainable Development in Small Caribbean islands*. The University of the West Indies, Centre for Environment and Development, Mona, Jamaica.
Pratt, C., Lindsay, K., Pearson, M. and Thomas, C. (2009) *The Wild Plants of Antigua and Barbuda: An Illustrated Field Guide to the Native and Naturalised Vascular Plants*. Environmental Awareness Group, Antigua.
Robson, J. and Robson, I. (1996) From shareholders to stakeholders: critical issues for tourism marketers. *Tourism Management* 17(7), 533–540.
Ross, T. (2010) *Report on Seabird, Land Bird and Lizard Surveys on Antigua's Offshore Islands, May-June 2010*. Report to the Environmental Awareness Group, Antigua.
Ross, T. (2011) *Report on Seabird, Land Bird and Lizard Surveys on Antigua's Offshore Islands, May-June 2011*. Report to the Environmental Awareness Group, Antigua.
Saarinen, J. (2006) Traditions of sustainability in tourism studies. *Annals of Tourism Research* 33(4), 1121–1140. DOI: 10.1016/j.annals.2006.06.007.

Sautter, E.T. and Leisen, B. (1999) Managing stakeholders: a tourism planning model. *Annals of Tourism Research* 26(2), 312–328.
Scheyvens, R. and Momsen, J. (2008) Tourism in small island states: From vulnerability to strengths. *Journal of Sustainable Tourism* 16(5), 491–510.
Scott, D., Simpson, M.C. and Sim, R. (2012) The vulnerability of Caribbean coastal tourism to scenarios of climate change related sea level rise. *Journal of Sustainable Tourism* 20(6), 883–898.
Seetanah, B. (2011) Accessing the dynamic economic impact of tourism for island economies. *Annals of Tourism Research* 38(1), 291–308.
Sheehan, L.R. and Ritchie, J.R.B. (2005) Destination stakeholders: exploring identity and salience. *Annals of Tourism Research* 32(3), 711–734.
Singh, N. (2008) Sustainable development in the Caribbean: from policies to people. *The Caribbean Environment* 1(1), 50–53.
Stevens, S. (ed.) (1997) *Conservation Through Cultural Survival: Indigenous Peoples and PAs*. Island Press, New York, NY, USA.
Streeten, P. (1993) The special problems of small countries. *World Development* 21(2), 197–202. DOI: 10.1016/0305-750X(93)90014-Z.
Strickland-Munroe, J. (2017) Resilience and PA tourism in Purnululu National Park: Understanding interactions with a focus on community benefits. In: Butler, R.W. (ed.) *Tourism and Resilience*. CAB International, Wallingford, UK, pp. 137–149.
Stronza, A. and Gordillo, J. (2008) Community views of ecotourism. *Annals of Tourism Research* 35(2), 448–468.
Theobald, W.F. (2005) *Global Tourism*. Elsevier Inc. Routledge, London, UK.
Timms, F.B. and Conway, D. (2012) Slow tourism at the Caribbean's geographical margins. *Tourism Geographies* 14(3), 396–418.
Timur, S. and Getz, D. (2008) A network perspective on managing stakeholders for sustainable urban tourism. *International Journal of Contemporary Hospitality Management* 20(4), 445–461.
Tkaczynski, A., Rundle-Thiele, S.R. and Beaumont, N. (2009) Segmentation: a tourism stakeholder view. *Tourism Management* 30, 169–175.
Vallega, A. (2007) The role of culture in island sustainable development. *Ocean and Coastal Management* 50, 279–300.
Waite, R., Kushner, B., Jungwiwattanaporn, M., Gray, E. and Burke, L. (2015) Use of coastal economic valuation in decision making in the Caribbean: enabling conditions and lessons learned. *Ecosystem Services* 11, 45–55.
Walker, T.B. (2017) *Contributions to Sustainable Tourism Development in Small Island Developing States (SIDS): The Citt slow approach for Antigua and Barbuda*. Doctoral dissertation. Ritsumeikan Research Repository.
Walker, T.B. (2019) Sustainable tourism and the role of festivals in the Caribbean – case of the St. Lucia jazz (& arts) festival. *Tourism Recreation Research* 44(2), 258–268.
Walker, T.B. (2020) A review of sustainability, tourism and the marketing opportunity for adopting the Cittàslow model in Pacific small islands. *Tourism Review International* 23(3-4), 99–111.
Walker, T.B. and Lee, T.J. (2019) Contributions to sustainable tourism in small islands: an analysis of the cittàslow movement. *Tourism Geographies* 23(3), 415–435.
Weaver, D.B. (1995) Alternative tourism in Montserrat. *Tourism Management* 16(8), 593–604.
Wise, S.P. (2014) Learning through experience: Non-implementation and the challenges of PA conservation in The Bahamas. *Marine Policy* 46(2014), 111–118.
Wondirad, A., Tolkach, D. and King, B. (2020) Stakeholder collaboration as a major factor for sustainable ecotourism development in developing countries. *Tourism Management* 78, 104024.
World Travel and Tourism Council [WTTC] (2020a) *Economic Impact Reports. Travel and Tourism Regional Performance 2019*. Available at: https://wttc.org/Research/Economic-Impact (accessed 3 February 2022).
World Travel and Tourism Council [WTTC] (2020b) *Travel and Tourism Global Economic Impact and Trends 2020*. Available at: https://wttc.org/Portals/0/Documents/Reports/2020/Global%20Economic%20Impact%20Trends%202020.pdf?ver=2021-02-25-183118-360 (accessed 3 February 2022).
World Travel and Tourism Council [WTTC] (2020c) *Antigua and Barbuda. 2020 Annual Research: Key Highlights*. Available at: https://wttc.org/Research/Economic-Impact/moduleId/1445/itemId/91/controller/DownloadRequest/action/QuickDownload (accessed 3 February 2022).

6 Contribution of Tourism to Sustainable Development: Samalayuca Dunes (Mexico)

Manuel Ramón González-Herrera*, Rosa Herminia Suárez-Chaparro and Karina Hernández-Casimiro
Autonomous University of Ciudad Juárez, Mexico

Abstract
In achieving the UN Sustainable Development Goals (SDGs), especially in protected natural areas (PNAs), it is necessary to promote the conservation of an area's natural and cultural heritage. The objective here is to potentiate a low-impact tourism model with the participation of the local community in the Samalayuca Dunes, Mexico, to achieve the SDGs. To this purpose, a participatory low-impact tourism strategy was built with stakeholders in gateway communities, based on three pillars: natural/cultural heritage; economic development; and social wellbeing. The strategy integrated six stages: a perception study on the development model; situational diagnosis of the Public Use Zone; study of natural and cultural resources for tourism use; tourism carrying capacity; sustainability criteria; and design of interpretive ecotourism trails. As a result, a sustainable tourism protocol, recommendations for a management and public use programme, an interpretation and environmental education programme, and a code of conduct for visitors were transferred to the community.

Keywords: tourism, sustainability, local participation, protected natural areas (PNAs), Samalayuca Dunes

6.1 Introduction

Tourism is a socioeconomic activity that generates wealth and employment, so it can play an important role in reaching the United Nations Sustainable Development Goals (SDGs) (García, 2019). Therefore, taking advantage of the benefits tourism provides must be key to achieving these goals and putting into practice the development agenda projected towards 2030 (United Nations World Tourism Organization, n.d.). Especially in protected natural areas (PNAs), it is necessary to promote the conservation of natural and cultural heritage, and at the same time favour the endogenous development of host communities.

The results of the participatory study of tourism development with sustainability criteria in the Samalayuca Public Use Zone, located south of Ciudad Juárez, Mexico, are presented; the study context includes the PNA Samalayuca Dunes and their area of influence, integrating 11 gateway localities of tourist interest of mainly local and regional scope. This project is important, because in this area a spontaneous, exogenous and non-participatory model of tourism development has been formed, which highlights the incompatibility of this model with the SDGs due to the obvious symptoms of unsustainability noticed by the local community and specialists from different academic, research

*Corresponding author: manglez04@yahoo.es

 Tourism Transformations in Protected Area Gateway Communities (eds S.L. Slocum *et al.*)
DOI: 10.1079/9781789249033.0006

and management institutions. This contradiction between the current model and the desired future of sustainable development constitutes the scientific problem assumed as the object of study.

Two scientific questions were formulated: How can tourism contribute to sustainability in Samalayuca, and what can be done to achieve it? and What should the role of gateway communities be in achieving this purpose? The research aim was that of potentiating a low-impact tourism model with the participation of the local community in the Samalayuca Dunes as a way to achieve the SDGs. As a possible answer, the following assumption was formulated: tourism could contribute to sustainability in Samalayuca, and gateway communities could play an important role in achieving this purpose, turning this destination into a more integrated, planned and participatory tourist space, with tangible benefits for the host communities and increased visitor satisfaction.

The main outcome of this research is the contribution of community stakeholders to the understanding of current tourism development, specifically within the realm of gateway communities, and the search for transformations towards sustainability. Regarding this, the specific outcomes are: situation assessment and identification of the attractions that characterize the vocational and tourist potential; perceived analysis of the possible contribution of tourism to the fulfilment of the SDGs and the current impacts that limit them; actions for the prevention and control of the negative impacts of tourism and the conformation of the desired low-impact tourism development model to achieve the SDGs; and projection of a participatory tourism development strategy to achieve the SDGs based on three pillars: natural and cultural heritage; economic community development; and social wellbeing. These outcomes are of value for local development as they offer an alternative option to traditional tourism development, which lack a community base and reflect a high background of negative impacts.

6.2 Literature Review

6.2.1 Implications of sustainable tourism theory

The process of formation and maturation of the sustainable development concept is complex and has been highly debated since its appearance in the late 1980s, due to different disciplinary interpretations (Calvente, 2007; Lohmann and Panosso, 2008; Orozco and Núñez, 2013; Vollero *et al.*, 2019). For this reason, its evolution has been associated with diverse approaches, including ideological, social construct, paradigm, and theoretical models, among others. The modern meaning of this term defends, although not without controversy, the integration of three dimensions related to development, which include the economic, social and environmental scope. During the last decades, numerous scientific studies have contributed to a better understanding and operationalization of the concept of tourism sustainability (Cabrini, 2014), clarifying the need to 'emphasize sustainable tourism as a strong paradigm and vision for an alternative development' (Perona and Molina, 2016, p. 527).

This research uses the definition of sustainable development contained in the Mexican legislation, accepted as an evaluable process through criteria and indicators of an environmental, economic and social nature, which tends to improve quality of life and productivity, based on the preservation of ecological balance, environmental protection and the use of natural resources (Ley General del Equilibrio Ecológico y la Protección al Ambiente, 2019). According to the General Tourism Law, sustainable tourism allows optimal use of natural resources suitable for tourism, helping to conserve them by respecting the sociocultural authenticity of the host communities, preserving the cultural, traditional and architectural values, and ensuring economic activities that improve living conditions by offering benefits, such as employment income, and other social services (Ley General de Turismo, 2019). Its implementation will allow benefits to the community along with the minimization of negative impacts, thus promoting community social development. Sustainable development recognizes that tourism should not negatively affect the quality of life of the community, but rather maximize benefits for cultural heritage and support the adoption of behaviour codes (Global Sustainable Tourism Council, 2012).

To understand the concept of sustainable tourism as a paradigm of development, different models and theoretical approaches have been presented, which can be postulated to improve comprehension of the contribution of tourism for achieving the SDGs. Among the selected models for the present study are: the '3P' model of sustainability: people, planet, profit; the triple bottom line model: economy, society and environment; and the circle of sustainability model: economics, ecology, politics and culture (Slaper and Hall, 2011; Thatcher, 2014; University of Wisconsin, n.d). The conceptual and methodological triangulation of these models is useful for clearly visualizing this study's objective since they are based on a holistic conceptualization of development and provide the basis for the incorporation of an integrative approach and the preventive management of impacts.

The assumed theoretical–conceptual implications permit the explanation (Fig. 6.1) that tourist value enhancement of the environmental attributes of the PNAs as attractions that characterize the tourist vocation can be favoured by incorporating the SDGs (input). Based on this premise, an analysis of planning scenarios was necessary, which started with the diagnosis of the problems of the current tourism model as a basis for the projection of the desired state and the corresponding strategy to achieve it (output), which is based on the three pillars of sustainability; this theoretical positioning implies the organization of community-based tourism, which incorporates local community participation as a methodological approach.

Among the instruments to promote sustainable tourism, the 2030 Agenda and the Strategic Framework for Sustainable Tourism in PNAs were selected. The United Nations

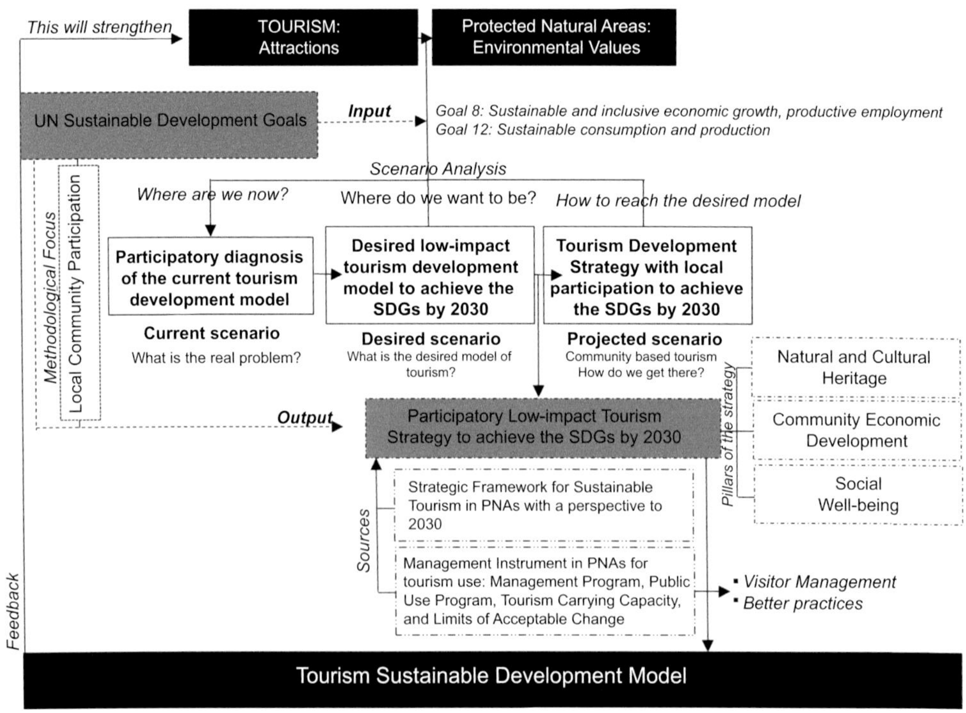

Fig. 6.1. Theoretical model (figure courtesy of the authors).

Agenda 2030 (2015) includes 17 SDGs with specific aims to be reached in the next 15 years, and goes further than its precursor, the Millennium Development Goals (MDGs), adopted in 2000, which should have been accomplished by 2015. Although both were intended to address the fundamental causes of poverty and the universal need for equality and human development for all (United Nations, 2015), the United Nations World Tourism Organization (2020) is in charge of promoting responsible, sustainable and accessible tourism and has in its sights the fulfilment of the 2030 Agenda and the SDGs with universal scope.

Of interest to this study is Goal 8, which proposes the promotion of sustained, inclusive and sustainable economic growth, full and productive employment and decent work for all, and Goal 12, aimed at guaranteeing sustainable consumption and production. A tourism sector that adopts such priorities could play a significant role in accelerating global transformation towards sustainability. In particular, Target 12.b indicates that developing and applying instruments that make it possible to closely monitor the effects of development, with a view to achieving sustainable tourism that creates jobs and promotes local culture and products, is advisable.

To achieve the SDGs in PNAs, it is necessary to introduce the concept of low impact environmental tourism. The definition of the National Commission for Protected Natural Areas in Mexico (CONANP, 2013) was used, which defines it as an environmentally responsible tourism modality, consisting of travelling or visiting relatively undisturbed natural spaces and cultural manifestations of the present and the past, for enjoyment, appreciation and study, through a process that promotes conservation, has low environmental and cultural impact and induces an active and socioeconomically beneficial involvement of local populations.

The Strategic Framework for Sustainable Tourism in PNAs (CONANP, 2018) allows the identification of the necessary components for the use of tourism and the minimum conditions of sustainability, generating a baseline of strategic information that facilitates decision-making. It recommends, among the instruments for the management and ordering of tourist use, the Management Program (p. 30); the Public Use Program (p. 31); the Study of Tourism Carrying Capacity (p. 31); and the Study of the Limit of Acceptable Change (p. 31). It also proposes mechanisms for the prevention and control of impacts, such as regulations for visitor management and the development of tourism initiatives based on 'better practices', for the benefit of local inhabitants inside, and in the influence areas, of the PNAs.

6.2.2 Role of communities in PNAs for achieving the SDGs

The integration of communities into tourism development is a basic requirement for participatory sustainability management (Clausen and Gyimóthy, 2016; Bridgewater and Rotherham, 2019). This approach is particularly necessary when the interests of the parties involved are opposed in the space–time dimension; primarily in cases where the expectations of tourists and the community are not convergent (Wang *et al.*, 2014; Crosby, 2015; Tseng *et al.*, 2019). It is advisable to favour the convergence of interests and expectations among the interested parties (Silva *et al.*, 2017), so that positive memorable experiences are promoted (Crosby, 2015). The participation of all the actors involved at the local level, specifically within the neighbouring gateway communities, constitutes an appropriate framework to potentiate the required synergies (Grybovych, 2012), including individuals, groups, public and private institutions, programmes, services and businesses (Rodríguez, 2015; Guccio *et al.*, 2017).

According to Skewes *et al.* (2012), it has become common to construct a critique of tourism according to the modernization pattern that it implies when it generates conflicts for the environment and minority societies. Current efforts to modify these stereotyped patterns mark the point where trends, such as community-based tourism, emerge, recognizing that tourism is supported by three key concepts, namely: community, conviviality and everyday life (Henríquez *et al.*, 2010). Such statements make tourism a more humane phenomenon, without necessarily losing its essence.

> Searching for alternatives to traditional models, locally-based tourism (LBT) arises as a strategy to promote development and environmental conservation for the sake of sustainable tourism; implies an organisational model of the activity focused on local control over the planning, management, and benefits of tourism … associated with this local control, different notions of community being implicit and becoming part of the LBT.
>
> (Ruiz-Ballesteros, 2015, p. 19)

It is recommended that various concepts 'such as community-based tourism, community ecotourism, and rural community tourism could be framed within the broader concept of LBT' (Ruiz-Ballesteros, 2015, p. 20), an approach used in this study. It takes the local community as the primary agent, which is why the 'main particularity of the LBT is its contribution to local development' (Ruiz-Ballesteros, 2015, pp. 40–41).

LBT is a sustainable option for economic and social development, together with a balanced distribution of benefits at the local level (Hamilton and Alexander, 2013; Perona and Molina, 2016). This approach is understood as 'a sustainable alternative to mass tourism, which has commercialized territories, displaced human populations and has transformed their spaces and living conditions' (Palomino *et al.*, 2016, p. 6). According to Palomino *et al.*, 2016, p. 8), LBT 'proposes a different use of the territory and its resources, both by those who provide the service and by users, as it involves valuation practices in situ, the recognition of cultural heritage of the communities and their interaction with them'.

Therefore, it is necessary to promote active and integrated community participation in planning and decision-making processes from the local level (Guccio *et al.*, 2017; Pulido and Pulido, 2017). Regarding tourism planning and management, Monterrubio (2009) introduces an important consideration by noting the relationship between tourism and the community dimension through community participation. The community tradition refers to the degree of participation exercised by the community when carrying out planning (Grybovych, 2012); this tradition suggests that the benefits provided by tourism favour the community. Of course, these communities will also be responsible for tourism planning and monitoring the projects, programmes or plans (Monterrubio, 2009).

In this way, the community that inhabits the destination should always be the most interested in obtaining knowledge about tourism in its coexistent space and should also enter the decision-making process, assuming the benefits and damages that may arise (Waligo *et al.*, 2015; Lin *et al.*, 2017). If the majority of the community does not agree with the tourist activity, the development of the sector will be conflictive and dysfunctional at some point in time (Valls, 2004), which is why the perceptions of the community, regarding tourism development, are of great value for the sustainable management of the destination.

6.3 Methodological Framework

This research is of an applied, explanatory-interpretive and transversal type, structured in two stages: diagnostic and projective. Qualitative, socially-interdisciplinary and deductive approaches were used, since the research spanned from the general theory of tourism sustainability to its implementation in the practical solution of a problem. Mixed information sources were used to obtain categorical and numerical data.

The orientation of the methodology was grounded on the principles of participatory strategic planning (Osácar, 2005). The strategic diagnostic phase was based on the study of perceptions of the current and desired tourism model, introducing the SDGs to the analysis of the vocation and tourism potentiality. In the strategic projection phase, the study of tourism potential (inventory and evaluation of attractions) was carried out, as well as the functional zoning. Based on previous results, discussion actions for the prevention and control of the negative impacts of tourism, and the projection of a participatory tourism development strategy to achieve the SDGs, were proposed.

The tourist attraction inventory was conducted via fieldwork by the research team with members of the gateway community. The information on each attraction was completed in situ and photos were taken, which were later evaluated by other local participants in this study (n=16). The perception of the tourist attractions was evaluated using three variables: tourist attraction (indirect observation and previous experience); attraction capacity (Likert scale); and affective-motivational relation (percentage).

The analysis units corresponded to stable groups of people living in Samalayuca with a sense of belonging to this area and to geographic units corresponding to the research polygon. Non-probability convenience sampling was applied, in which the sample was made up of locals who participated in focus groups (n=8). Analytical-synthetic, hermeneutic-interpretive modelling, and geospatial theoretical methods were used, as well as empirical methods, such as participant observation, document analysis and group studies through discussion. The data collection tools included observation guides, focus group script, checklists and inventory tables. The types of non-comparative measurement scales used included: graphical data measurement, Likert, semantic differential and data analysis matrices. The study of the realities of the participants was based on their testimonies, anecdotes, comments, opinions and examples.

Research development was based on the stages that make up the research method (Fig. 6.2), acknowledging scientific observation as a starting point to arrive at new scientific knowledge. The stages and research tasks were: documentary research; characterization of the research context; diagnostic study; development of solution proposals; and preparation of conclusions and recommendations. The development of strategies for the adoption of the research included the dissemination of the conclusions among the community stakeholders, the use of the results by decision-makers, and the commitment to feed back the conclusions of the research to the community and inform them of the policies and the best practices to implement.

6.4 Situation Assessment

The Samalayuca Dunes are located in northern Mexico, within the Chihuahua Desert, and are considered to be the dunes with the greatest biodiversity in North America. This space is represented by a system of wandering dunes of fine

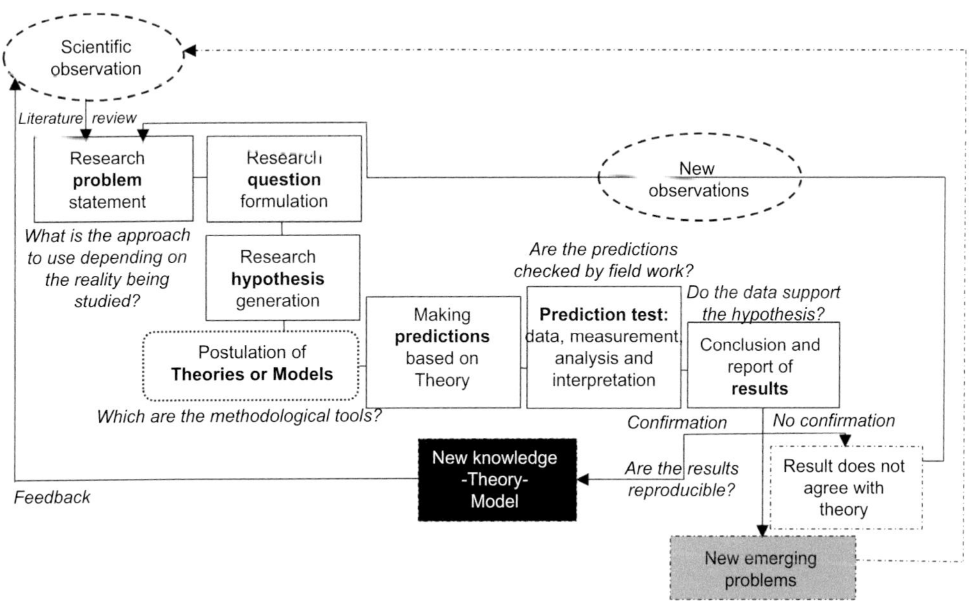

Fig. 6.2. Stages of the scientific method (figure courtesy of the authors).

sand composed of silica dioxide, recognized as the highest dunes in Mexico, so it constitutes one of the most important representations of dunes nationwide and one of the most extraordinary dune areas on the planet. This territory belongs to the hydrological region of closed basins in the north, in arid conditions, without superficial water flows but with important groundwater reserves. It is characterized by significant endemisms – the habitat contains approximately 240 species of plants and 150 species of fauna. It also contains more than 50 archaeological sites and is the location of the settlement of rural communities.

The origin of the human population is related to the establishment of cattle ranches favoured by the presence of wells or springs. The official foundation of the town dates from 1880 as a result of the Mexico-Juárez Railway. The main town, and other rural settlements, are located very close to the north–south road axis that makes up Federal Highway 45, otherwise known as the Pan-American Highway.

The main economic activities are stockbreeding, agriculture, charcoal making, mining, beekeeping and tourism, which represent the main sources of employment, although most of the local population do not work in tourism. According to the participants, these activities provide the main sources of income; in this regard, one of the *ejidatarios*[1] reported '... I am dedicating myself to beekeeping, which I like to do not as a job but as a hobby ... which in addition to contributing to the economy provides benefits to health and the environment'. Another participant added that the people of the village are not wealthy – things have to be done according to the characteristics of the place. For example, they build cabins with materials from the area. In the vicinity of Samalayuca, industrial activity is also represented, as a cement plant and two thermoelectric plants are located there.

The geospatial structure of the environment, determined by the uneven combination of its components, and the different types and levels of anthropic assimilation, allowed the delimitation of four geoenvironmental units (Fig. 6.3).

In 2009, part of the Samalayuca Dunes was declared a PNA under the category of Protection of Flora and Fauna. Its general purpose is the conservation, protection, restoration and management of the habitat of wild flora and fauna species. Its scenic landscape is of great beauty, and it is possible to carry out recreational, educational and research activities, as well as the sustainable use of existing natural resources, and to promote the development of this space, although tourism is not explicitly stated within the Samalayuca Dunes designation policy. As

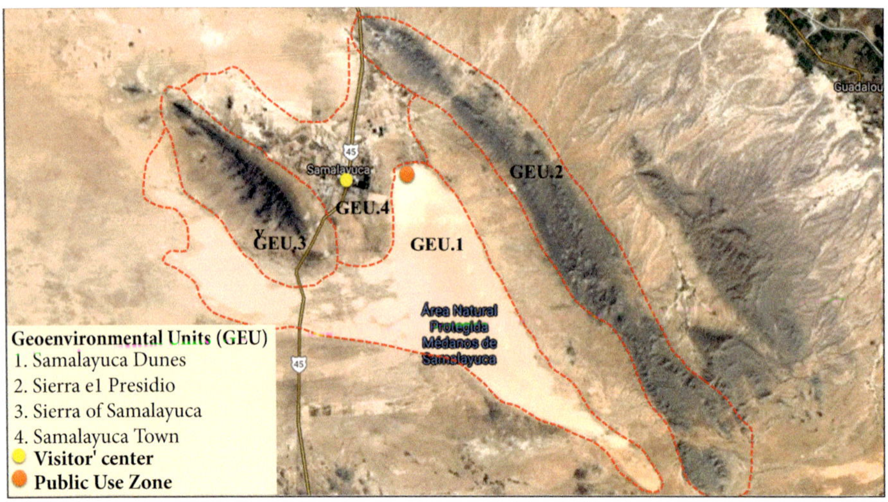

Fig. 6.3. Geoenvironmental units (image courtesy of the authors/ with www.google.com/maps/@31.3240644,-106.4653535,49178m/data=!3m1!1e3).

part of its purpose, the management programme establishes the preservation of natural environments, archaeological, historical and artistic vestiges, as well as tourist areas and others of importance for the recreation, culture and identity of the Chihuahuan Desert (Comisión Nacional de Áreas Naturales Protegidas, CONANP, 2013).

6.4.1 Attractions that characterize tourist potential

Through fieldwork with community members, an inventory of tourist attractions was carried out. It was found that the attractions comprise natural, rural, archaeological, cultural and artificial resources structured at 11 sites. Based on the inventory, a participant evaluation of the potential of tourist attractions was conducted according to their capacity to attract tourists and the affective-motivational object–subject relationship, identifying the most valued attractions (Table 6.1). The geological/geomorphological formations corresponding to the dunes were evaluated in the international hierarchy due to the ecological, aesthetic, cognitive and recreational values that they possess, which attract a flow of international tourists.

6.4.2 Impact perception

A brief environmental history associated with anthropic activities carried out within the natural environment was reconstructed. Access to the dunes is relatively recent, and according to locals, it was related to filming in their landscape. A participant commented:

> When offering scenes of interest to world-famous film directors and actors in the 1980s, the film's recording team ... opened the current access, before that, for us the dunes were only like a dream when we saw the sand hills from afar. In the absence of access, it was not easy for people to get there because the walks were too long. If someone needed to go it was only to open spaces such as drinking fountains for their livestock.

Later, the tourist boom in the dunes put an end to raising cattle, goats or sheep, and the vocation of the locals was modified. They adapted their agricultural activities (*nogalera*), although it is still possible to find livestock activity in *Ejido Ojo de la Casa*.

Such growth in tourism has not been planned, nor oriented towards sustainability. One participant pointed out that, in terms of tourist activity, the visitors must be taught to take care of the environment. The park needs to make the origin of the place known, not only to tourists, but also to the community, since residents have seen a sudden increase in tourism, and they are unaware of how to take best advantage of it. The participants recognized the importance of tourism as an economic activity for the site. The main positive impacts perceived by the participants were related to the revaluation and putting into use of the dunes, the generation of income and sources of employment, the role of CONANP in the conservation and protection of the site, and the regulation of visitor flows through the entrance fee.

The main perceived factors that bring about negative impacts are: insufficient planning; illegal activities; conflicts of interest; lack of environmental culture; and lack of control and systematic monitoring. Among the perceived environmental impacts that affect progress towards sustainability were: extraction of sand; mesquite felling; changes in land use; accumulation of tyres and other contaminants; extraction and sale of the desert rose (an intricate rose-like formation of crystal clusters); illegal extraction of native fauna and introduction of foreign species; as well as the practice of illegal mass tourism activities (uncontrolled overtourism) and inherited impacts from earlier times.

It was identified that the main negative impact felt by the locals is environmental degradation and contamination. In the opinion of a participant, 'since the tourist boom began to be generated in the dunes with people who no longer went to the Juárez Valley, the tourist activity brought activities that give a bad image and cannot be controlled, giving the appearance of a canteen, brothel, drug addiction, and dirt'. Another participant mentioned that one of the problems generated by tourist activity is pollution, specifically garbage. Another *ejidatario* mentioned, as a problem, the consumption of

Table 6.1. Perception of tourist attractions.

Tourist attraction	Attraction capacity	Affective-motivational relation
 **Fig. 6.4.** Samalayuca Dunes (photo courtesy of the authors).	Functional value: 5 Aesthetic/scenic value: 5 Environmental value: 5 Historic value: 5 (natural) Symbolic value: 5 Cognitive value: 5 Authenticity: 5 Accessibility: 5 Infrastructure and services: 4	Spectacular: 95% Very nice: 5% Nice: – Moderately pleasant: – Unpleasant: –
 **Fig. 6.5.** Exhacienda (photo courtesy of the authors).	Functional value: 4 Aesthetic/scenic value: 5 Environmental value: 4 Historic value: 5 Symbolic value: 5 Cognitive value: 5 Authenticity: 5 Accessibility: 4 Infrastructure and services: 3	Spectacular: 80% Very nice: 20% Nice: – Moderately pleasant: – Unpleasant: –
 **Fig. 6.6.** Sierra of Samalayuca (photo courtesy of the authors).	Functional value: 5 Aesthetic/scenic value: 5 Environmental value: 5 Historic value: 5 Symbolic value: 5 Cognitive value: 5 Authenticity: 5 Accessibility: 4 Infrastructure and services: 3	Spectacular: 85% Very nice: 15% Nice: – Moderately pleasant: – Unpleasant: –

Continued

Table 6.1. Continued

Tourist attraction	Attraction capacity	Affective-motivational relation
 **Fig. 6.7.** Petroglyph (photo courtesy of the authors).	Functional value: 5 Aesthetic/scenic value: 5 Environmental value: 5 Historic value: 5 Symbolic value: 5 Cognitive value: 5 Authenticity: 5 Accessibility: 4 Infrastructure and services: 3	Spectacular: 85% Very nice: 15% Nice: – Moderately pleasant: – Unpleasant: –
 **Fig. 6.8.** Desert rose (photo courtesy of the authors).	Functional value: 5 Aesthetic/scenic value: 5 Environmental value: 5 Historic value: 5 Symbolic value: 5 Cognitive value: 5 Authenticity: 5 Accessibility: 4 Infrastructure and services: 3	Spectacular: 90% Very nice: 10% Nice: – Moderately pleasant: – Unpleasant: –

alcohol and the use of motorized vehicles within the area, which can cause accidents to families who visit the dunes.

Another problem of interest for the community is the environmental degradation of the dunes. A participant pointed out:

> Historically we lived from the exploitation of sands that we sold, which represented extra income for us. It was before knowing the importance of keeping the resources in place. When CONANP arrived, the regulations changed so that the sand remained in the area. This is how many times we think only of the immediate economic benefit. In the case of *El Vergel* it is better that the roses of the desert are preserved in the place, so that they are always there and people come to see them.

The indiscriminate looting of mesquite firewood is another important problem. This tree has been replaced with *plumerillo*, which has ornamental value, although it is a foreign element to this place and consumes a lot of water.

The lack of first aid and security services for tourists was identified, and the health of tourists is not assured if an incident occurs: 'in case a visitor is run over in the dunes, where do we take them to receive medical attention? However, we do have a permanent ambulance in Samalayuca with equipment and trained personnel'. It was mentioned that *Dunas Campestre* has a fixed-mobile unit in high seasons. It was, therefore, proposed that the Union of Recreational Centres forms a committee responsible for a medical service, in charge of the wellbeing of tourists and visitors, sustaining this project through monetary and individual contributions. It was stated that CONANP is in charge of everything related to tourism and is the arrowhead to open up to tourism, therefore CONANP is the only agency available to take responsibility.

However, it was made known that the people who visit the dunes feel safe, and it is the responsibility of the local residents to continue maintaining that feeling of security. Currently the place is quiet without criminal problems, despite the fact that in the town there is no police station and just three agents to serve the four *ejidos*. It was reiterated that the town lacks medical care and that a positive change has been seen because of charging entry to the dunes.

Among the management decisions best valued by locals, in response to the situation of territorial deterioration that has occurred up to now, we should highlight:

- The regulation of the number of people entering the PNA – a resident commented, 'It is better to have the area controlled since the fewer people entering the site, the better the existing resources can be taken care of.'
- The charge of 100 pesos (less than US$5) as a fee to enter the area, intended for the conservation of the natural area – a participant commented, 'at the beginning we were afraid that our resources would decrease. With the change we have the same income, the area is more controlled, that is, fewer people with the same income and we have seen more control related to the garbage and bad vices.'
- Planning studies of the territory – a participant commented, 'Work is already under way on the carrying capacity study based on the limit of acceptable change.'

6.4.3 Prevention and control of negative impacts

Searching for transformations towards sustainability, the need for environmental education for residents and visitors was highlighted. A participant affirmed, 'among the things that we should focus on is tourism awareness and care for the environment, with support from the university and CONANP'. Another participant expressed the need to place more containers to deposit garbage, even though in the last year there has been a decrease in garbage, which may be due to the administrators and the emphasis placed on this issue. More signage for environmental awareness, with the use of images, was suggested as they have a greater impact because images speak for themselves.

Emphasis within the community on the importance of taking care of the environment was also commented upon. Since the growth of tourism and the environment are not in conflict, integration is sought, in order to form a group that really values the characteristics of the site. A participant mentioned, 'in my family we have managed the use of resources in a similar way, however, with the passage of time, my father and I have considered making a tourist corridor so that they can observe the desert roses on site and carry out accompanied tours, and place camping areas as a complement for the visitors who go to the dunes'.

One of the most debated topics was the conflict between extractive activities and the activities of those who wish to preserve natural values. A participant affirmed:

> I consider that the exploitation of the desert rose should take place in a balanced way. There are pieces that can be used to have an economic income, while other stones can be used for souvenirs, and others kept in the place as part of a gallery-museum. In this way, there would be a balance in the area and the identity would not be lost. It is necessary to know the stone capacity that we have, since up to now, it has not been known how much there is and if it could allow extraction for greater use. Lately

> very small stones have been obtained in the extraction which we use for costume jewellery. My family has a more direct connection with other commercial activities, and we have remained outside the tourist activity, perhaps this is due to a lack of knowledge and advice, however, we have interest and the possibility of integrating ourselves into tourism.

The immediate benefit is seen in the extraction and commercialization of the desert rose. The idea of an interpretive geotourism trail is evaluated, where visitors pay for an experience with this attraction in situ, which could produce economic and environmental benefits. According to the recommendations of the participants, the use of desert roses could be to expose the rocks and explain their characteristics, and at the same time they would be able to rescue mesquite and plant these typical trees of the area.

6.4.4 Tourism contribution to the SDGs

Regarding Objective 8, the locals recognized that the resources they have are an important productive factor for economic growth and a source of jobs; in the same way they acknowledged, with respect to Objective 12, that the production of goods and services associated with heritage values could potentiate better local development. Such is the case in the opinion of one participant when acknowledging that 'among the activities carried out is sandboarding, which attracts many visitors and even champion athletes from Brazil ... for which reason I consider that a better use of this activity should be given for the welfare of the community'. These perceptions coincide with the reported findings in the reviewed literature (Crosby, 2015; Clausen and Gyimóthy, 2016) and with the theoretical and methodological approach established in Fig. 6.1.

The importance of the site and its international value was recognized, which participants related to the motivation it awakens among visitors from different countries. A participant stated that:

> Although there is a lack of knowledge in the community to understand this importance, it is true that the actions by CONANP, among which are the teaching of environmental courses, mainly to children who are a fundamental part of the project, have allowed us to recognize the resources of this place, as well as care for the environment.

Another participant referred to the international hierarchy of attractiveness by saying, 'I have lived most of my life here in the town, as my grandparents were among the founders, so I think the international importance that dunes have should be recognized.' Similarly, they reported the arrival of different international tourists, and one participant commented that:

> As I was in charge of caring for the dunes for six years, I have seen people from various Mexican states and also foreigners ... with the exclusive interest of visiting the dunes. In the talks I had with a Spaniard, he told me that he had had great interest in coming to these dunes, even though he lived closer to the Sahara.

6.4.5 Desired tourism development model for achieving the SDGs

Regarding future projections, an integrated tourism project should be formed, on a reduced scale, that is neither massive nor invasive. The idea of developing large projects, such as hotels, was not supported by the community, as it would provide jobs for outsiders, which would not represent benefits to the community. A participant said, 'we do not want megaprojects, and we are grateful for decreases in the number of visitors, which has allowed greater conservation'. Another suggested: 'A museum where nature is appreciated. My family has dedicated itself for years to the extraction of the desert rose ... Once they took me to analyse a rose, it was there where I understood the treasure it represents, I realised the effects that extraction will bring.' There is an ideal for the future of which tourism is a part, indicating 'one of the dreams for which we are working is to turn Samalayuca into a magical town'.

The desired model should be endogenous and participatory, and include diversified territorial development, which is supported by the potential of the attractions of each site, and through which specialized products and representative activities can be offered. This idea is based on the fact that

> it is necessary to take advantage of the benefits of each *ejido; Ojo de la Casa* with petroglyphs, *El Vergel* with desert roses, *Villa Luz* with the dunes, and Samalayuca with agricultural activities, beekeeping, among others ..., offering the visitor activities in the different places, the tourist would get more involved with the community.

The importance of active participation in the project was accepted for preserving the cultural and historical heritage of each site and those recreational activities that can be carried out, which was evidenced by the introduction that *El Vergel* is having to tourism. It is advisable to encourage and build trust among residents, so that they seek support and advice, providing them with the confidence to undertake a project, and in this way, the financial resources can be obtained. They are the owners of their resources and have the ability and talent to make the most of them; all they need is guidance and advice so they can make the best decisions.

The perceptions and comments of the participants confirm the lack of knowledge in the community regarding tourist activity, although it is true that they comprehended its importance, as well as the actions carried out by CONANP in favour of sustainable tourism. A notable sense of belonging of the community stakeholders with respect to the PNA, and the commitment to contribute to its care, was corroborated, an aspect that stands out as one of the first feelings which emerged among the participants. It was possible to observe, mainly among the women, a feeling of love and positive attitude towards the care and preservation of the area and the will to integrate the community into tourism.

6.5 Tourism Development Strategy for Achieving the SDGs

Taking, as a premise, that the current scenario is characterized by unsustainable conditions, with the features of oversized mass tourism, movement towards a tourism sustainability scenario based on the SDGs was proposed. Based on the reviewed literature and the statements presented in Fig. 6.1, the establishment of a model of tourism development committed to six attributes was decided: attractiveness; security; inclusiveness; reliability; competitiveness; and community wellbeing. A participatory sustainable tourism strategy was built with community stakeholders based on three pillars: natural and cultural heritage including protection and conservation, reduction of environmental impacts, and environmental education and interpretation; economic community development with locally-based attractive products, decent employment and local economic benefit; and social wellbeing incorporating respect for local identity, self-management and local security, and inclusion and community participation (Fig. 6.9).

Based on the construction of the model, 35 strategic actions were identified to promote minimum conditions of sustainability. These strategic actions were integrated in three stages as a part of a protocol to achieve sustainability (Table 6.2). The time horizons will allow the transition from priority to more long-term actions, according to the type of problem and the feasibility of execution.

A tourist zoning scheme was built for the Public Use Zone with sustainability criteria (Table 6.3), since this space requires the highest levels of protection and management. Different areas of tourist use were identified with the participation of locals, projecting the predominance of uses in the extensive category and restricting the intensive use of these areas. The zoning took into consideration the three theoretical frameworks of sustainability presented in this research – the economic, social and environment aspects – and the theoretical model presented in Fig. 6.1.

6.6 Conclusion

Our conclusion is to emphasize the need to sensitize the different community stakeholders to the understanding and implementation of the concept of sustainable tourism as a paradigm and development model for achieving the SDGs, recognizing the value of the instruments implemented in a participatory manner for sustainable tourism management,

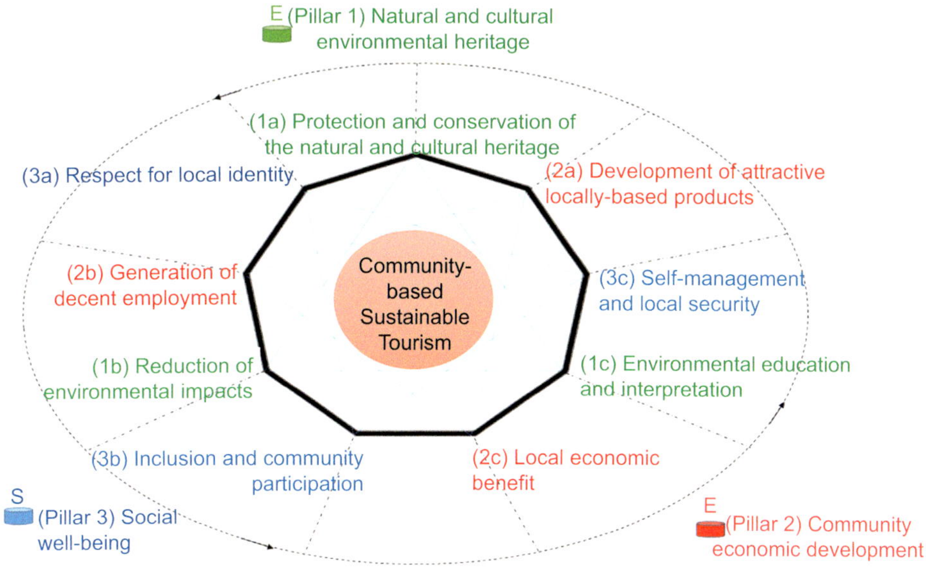

Fig. 6.9. Sustainable tourism development model (figure courtesy of the authors).

and the minimization of the impacts generated by tourism. It was found that the contribution of tourism to sustainability in Samalayuca is feasible if the role of local gateway communities continues to be strengthened as part of the planning and management activities of the territory, with integrated criteria of economic growth, protection of the natural/cultural heritage and social wellbeing. Such an approach requires a conceptual and operational rethinking of current tourism to enhance compliance with the SDGs by 2030, a horizon towards which a multifunctional, more integrated, planned and participatory space is projected with the purpose of promoting a more attractive, experiential and safe offer, which in turn promotes high levels of visitor satisfaction.

The favourable contribution of local participants to the diagnosis and strategic projection of sustainable tourism development was corroborated by the locals. It was based on their knowledge and wishes related to the protection of their natural and cultural spaces, as well as the will to place themselves 'in' tourism as a way to obtain economic and social benefits. The community views coincided with the main findings reported in the reviewed literature, especially in relation to the steps local inhabitants should follow for more sustainable tourism.

The limitations of the study were the small sample size, the lack of updated databases and poor scientific production on the subject at the local level. In the next stages of research, it will be necessary to promote the design of an interpretive trail, a tourist code of responsible behaviour, education and environmental interpretation manuals, as well as to favour the application of tourism marketing and commercialization with sustainability criteria. The incorporation of standards of sustainable tourism to the operation and management of tourism, the use of financing opportunities and development of sustainable projects has been recognised by the community stakeholders as decisive for future sustainable development. For the previously recommended reasons, new research questions, a literature review and predictions should be elaborated, as indicated in Fig. 6.2.

Table 6.2. Sustainable tourism protocol.

Actions for sustainable tourism development		
(A) First stage	(B) Second stage	(C) Third stage
Inventory of attraction sites	Adaptation to climate change	Regulations for tourism service providers
Pressure and intensity of use according to carrying capacity (overtourism)	Information for all stakeholders	Stakeholder participation
Care of natural and cultural heritage sites	Fair and equitable employment and training opportunities for women, youths and minorities	Protection of local access to genetic resources within the PNA
Protection of local cultures and traditional knowledge	Tourist awareness of all stakeholders	Support for local entrepreneurs and fair trade
Protection of wildlife	Training for tourism service providers	Integration of economic information on tourist activities
Security for all stakeholders	Study and monitoring of local community opinion	Use of local products and services based on their ecological, social and cultural conditions
Water and energy consumption	Provision of information and interpretation tools to the visitor	Holiday operations for visitors' orientation
Waste management	Low impact transportation	Monitoring of environmental costs throughout the tourism product chain
Controlled visitor management	Inventory and knowledge about risks	Articulation of tourism service providers in value chains
Visitors' code of responsible behaviour	Visitor satisfaction	Investment in tourism promotion and dissemination
Environmental monitoring and inspection	Access for all stakeholders	Investment in infrastructure and equipment with approaches to bioarchitecture and environmental technologies
Local community satisfaction	Crisis and emergency preparedness and response	

Table 6.3. Tourist zoning scheme.

Functional zone	Infrastructure and equipment	Tourist recreational/support and related services	Tourist activities
Administrative zone	Offices, bathrooms, multipurpose room, scientific research and monitoring centre	Drinking water supply, electricity supply, mobile network, internet access, toilets	–
Common use zone	Entrances and exits, parking, esplanade (plaza), ramps and walkways with handrails, viewpoints and observation tower, information module, artificial shadows, food processing, restaurant, guardhouse, first aid, dry toilets, garbage deposits, signage, environmental education classroom	Supply of drinking water, electricity supply, food and beverages, lockers, information (brochures, guides, interpretive signs, preventive, restrictive and informative signs, others), mobile network, internet access, toilets, equipment rental/sale, security, ground transportation, adventure/ nature operator, sledding sales	Observation of geological forms and relief (desert roses, dunes, etc.), observation of flora and fauna, observation of the landscape, observation of the sky (astronomical), souvenirs and crafts purchase
Camping zone	Tourism signage	Drinking water supply, electricity supply, accommodation (camping), toilets, mobile network, internet access, security	Camping, social activities, participatory games
Family use zone	Ramps and walkways with handrail	Mobile network, internet access, security	Sandboarding, observation of geological forms and relief (desert roses, dunes, etc.), observation of flora and fauna, observation of the landscape, observation of the sky (astronomical), yoga and meditation, filming and photography, concerts and events, screenings of movies
Children's play zone	Tourism signage	Safety, assistance and guidance, specialized guides	Children's sandboarding, educational games, observation of geological forms and relief (desert roses, dunes, etc.), observation of flora and fauna, observation of the landscape, observation of the sky (astronomical)

Continued

Table 6.3. Continued

Functional zone	Infrastructure and equipment	Tourist recreational/support and related services	Tourist activities
Extreme activities zone	Tracks, stands, signage and tourism signage	Mobile network, internet access, security, assistance, specialized guides, equipment	Sandboarding, motocross, cycling, running, observation of geological forms and relief (desert roses, dunes, etc.), observation of flora and fauna, observation of the landscape, observation of the sky (astronomical)
Environmental interpretation zone	Interpretation centre, interpretive trail with stations and interpretive stops, ramps and walkways with handrails, tourism signage, QR codes to inform about flora, fauna, etc.	Mobile network, internet access, interpretive guides, security, environmental education and interpretation services, biosecurity of nature destinations (viruses)	Interpretive hiking, observation of geological and relief forms (desert roses, dunes, etc.), observation of flora and fauna, observation of the landscape, observation of the sky (astronomical), environmental education workshops, guided tours

Note

[1] *Ejido*: land possession system in Mexico consisting of granting land to a group of people for their exploitation. *Ejidatario*: agrarian subject member of the *ejidal* nucleus. Source: Chamber of Deputies of the H. Congress of the Unio, *Agrarian Law*. Title Three, Chapter I, Section One and Section Two. Last Reform DOF 25-06-2018. Available at: http://wwwdiputadosgobmx/LeyesBiblio/pdf/13_250618pdf (accessed 4 February 2022).

References

Bridgewater, P. and Rotherham, I.D. (2019) A critical perspective on the concept of biocultural diversity and its emerging role in nature and heritage conservation. *British Ecological Society* 1(3), 291–304. DOI: 10.1002/pan3.10040.

Cabrini, L. (2014) *Trends in Sustainable Tourism and the Role of the Global Council (GSTC)*. Available at: http://docplayer.es/2312294-Tendencias-del-turismo-sostenible-y-el-rol-del-consejo-global-gstc.html (accessed 4 February 2022).

Calvente, A. (2007) El concepto moderno de sustentabilidad. *Socioecología y Desarrollo Sustentable* 1, 1–7.

Clausen, H. and Gyimóthy, S. (2016) Seizing community participation in sustainable development: Pueblos Mágicos of Mexico. *Journal of Cleaner Production* 111B(2016), 318–326.

CONANP (Comisión Nacional de Áreas Naturales Protegidas) (2013) *Programa de Manejo Área de Protección de Flora y Fauna Médanos de Samalayuca*. Available at: https://simec.conanp.gob.mx/pdf_libro_pm/33_libro_pm.pdf (accessed 4 February 2022).

CONANP (Comisión Nacional de Áreas Naturales Protegidas) (2018) *Marco Estratégico de Turismo Sustentable en Áreas Naturales Protegidas de México*. Secretaría de Medio Ambiente y Recursos Naturales, México. Primera edición.

Crosby, A. (2015) *Convergence of Interests Between Tourists, Locals and Territory*. EFEVERDE. Available at: http://www.efeverde.com/opinion/convergencia-de-intereses-entre-turistas-locales-y-las-areas-naturales

García, R. (2019) *El Turismo y los Objetivos de Desarrollo Sostenible*. AprendedeTurismo.org. Available at: https://www.aprendedeturismo.org/el-turismo-y-los-objetivos-de-desarrollo-sostenible-de-naciones-unidas/#:~:text=El%20turismo%20genera%20riqueza%20y,los%20Objetivos%20de%20Desarrollo%20Sostenible.&text=Seg%C3%BAn%20la%20Organizaci%C3%B3n%20Mundial%20del,los%20turistas%20mundiales%20en%202030 (accessed 4 February 2022).

Global Sustainable Tourism Council (2012) *Global Sustainable Tourism Criteria for Hotels and Tour Operators*, 2. Madrid, Spain. Available at: https://www.gstcouncil.org/en/gstc-criteria/criteria-for-hotels-and-tour-operators.html (accessed 4 February 2022).

Grybovych, O. (2012) Designing a qualitative multi-case research study to examine participatory community tourism planning practices. In: Hyde, K.F., Ryan, C. and Woodside, A.G. (eds) *Advances in Culture, Tourism and Hospitality Research: Field Guide to Case Study Research in Tourism, Hospitality and Leisure*. Emerald Group Publishing Limited, Bingley, UK, pp. 501–520.

Guccio, C., Lisi, D., Martorana, M. and Mignosa, A. (2017) On the role of cultural participation in tourism destination performance: an assessment using robust conditional efficiency approach. *Journal of Cultural Economics* 41(2), 129–154. DOI: 10.1007/s10824-017-9295-z.

Hamilton, K. and Alexander, M. (2013) Organic community tourism: a co-created approach. *Annals of Tourism Research* 42(1), 169–190. DOI: 10.1016/j.annals.2013.01.015.

Henríquez, C., Zechner, T.C. and Cioce Sampaio, C.A. (2010) Turismo y sus interacciones en las Transformaciones del Espacio Rural. *Revista Austral de Ciencias Sociales* 18, 21–31. DOI: 10.4206/rev.austral.cienc.soc.2010.n18-02.

Ley General de Turismo (2019) *Nueva Ley publicada en el Diario Oficial de la Federación el 17 de junio de* 2009. Available at: http://www.diputados.gob.mx/LeyesBiblio/pdf/LGT_310719.pdf (accessed 4 February 2022).

Lin, Z., Chen, Y. and Filieri, R. (2017) Resident-tourist value co-creation: The role of residents' perceived tourism impacts and life satisfaction. *Tourism Management* 61(1), 436–442. DOI: 10.1016/j.tourman.2017.02.013.

Lohmann, G. and Panosso, A. (2008) *Tourism Theory*. Editora Aleph, Sao Paulo, Brazil.

Monterrubio, J. (2009) Comunidad receptora: Elemento esencial en la gestión turística. *Gestión Turística* 11, 101–111.

Orozco, J. and Núñez, P. (2013) Las teorías del desarrollo: en el análisis del turismo sustentable. *InterSedes: Revista de las Sedes Regionales* XIV(27), 144–167.

Osácar, E. (2005) Planeación estratégica. In: Osácar, E., López, E. and Vilaginés, A. (eds) *Instrumento de Gestión Estratégica del Turismo*. CETT-UB, Barcelona, Spain, pp. 1–36.

Palomino, B., Zamora, J. and López, G. (2016) Community tourism in the Sierra Norte de Oaxaca: perspective from institutions and governance in Indigenous territories. *The Sustainable Periplo* 30, 6–37.

Perona, E. and Molina, E. (2016) Evolution and stages of tourism development in the province of Córdoba, Argentina: How far is the sustainable tourism ideal? *Notebooks of Tourism* 37, 527–530.

Pulido, J. and Pulido, M. (2017) Proposal for an indicators system of tourism governance at tourism destination level. *Social Indicators Research* 137(2), 1–49.

Rodríguez, M. (2015) The praxis of community social work at the local level: citizenship, participation, empowerment. *AIS, Actions and Social Investigations* 35, 7–29.

Ruiz-Ballesteros, E. (2015) Turismo de Base Local y Comunidad. *¿Una Vinculación Oportuna? Revista Andaluza de Antropología* 8, 19–44.

Silva, J., Agapito, D. and Pinto, P. (2017) Measuring the quality of destinations. In: Scott, N., De Martino, M. and Van Niekerk, M. (eds) *Bridging Tourism Theory and Practice: Knowledge Transfer to and Within Tourism*. Emerald Publishing Limited, Bingley, UK, pp. 221–237.

Skewes, J., Henríquez, C. and Vera, M. (2012) Turismo comunitario o de base comunitaria: una experiencia alternativa de hospitalidad vivida en el Mundo Mapuche. Tralcao Sur de Chile. *Revista de Cultura e Turismo* 6(2), 73–85.

Slaper, T. and Hall, T. (2011) The triple bottom line: What is it and how does it work? *IBR* 86(1), 4–8.

Thatcher, A. (2014) Theoretical definitions and models of sustainable development that apply to human factors and ergonomics. *11th International Symposium on Human Factors in Organisational Design and Management (ODAM 2014) CPH Conference*, Copenhagen, Denmark. Volume: 12.

Tseng, M.-L., Lin, C., Remen Lin, C.-W., Wu, K.-J. and Sriphon, T. (2019) Ecotourism development in Thailand: community participation leads to the value of attractions using linguistic preferences. *Journal of Cleaner Production* 231(3), 1319–1329. DOI: 10.1016/j.jclepro.2019.05.305.
United Nations (2015) *Objetivos de Desarrollo Sostenible. 17 objetivos para transformar nuestro mundo.* Available at: http://www.un.org/sustainabledevelopment/es/objetivos-de-desarrollo-sostenible/ (accessed 4 February 2022).
United Nations Agenda 2030 (2015) *Objetivos de Desarrollo Sostenible.* Available at: https://www.un.org/sustainabledevelopment/es/2015/09/la-asamblea-general-adopta-la-agenda-2030-para-el-desarrollo-sostenible/ (accessed 9 February 2022).
United Nations World Tourism Organization (n.d.) *El Turismo en la Agenda 2030.* Available at: https://www.unwto.org/es/turismo-agenda-2030 (accessed 4 February 2022).
United Nations World Tourism Organization (2020) *Turismo por los ODS.* Available at: https://www.unwto.org/es/tourismo-por-los-ods (accessed 4 February 2022).
Valls, F. (2004) *Gestión de destinos turísticos sostenibles.* Gestión 2000, Barcelona. España.
Vollero, A., Siano, A. and Volpe, M. (2019) A systems perspective for conceptualizing sustainability in long-lived family businesses. *System Research and Behavioral Science* 36(1), 111–127.
Waligo, V., Clarke, J. and Hawkins, R. (2015) Embedding stakeholders in sustainable tourism strategies. *Annals of Tourism Research* 55(1), 90–93. DOI: 10.1016/j.annals.2015.09.002.
Wang, S., Zhou, L., Lee, S. and King, C. (2014) Analysis of residents' social identity, tourism engagement, and propensity for tourism advocacy. *Advances in Hospitality and Leisure* 10, 109–129.

7 Community Engagement with Tourism Management in Small Atlantic islands

Maria A. Ventura*, Ana C. Costa and Andrea Z. Botelho
CIBIO, Centro de Investigação em Biodiversidade e Recursos Genéticos, InBIO Laboratório Associado - Pólo dos Açores; Faculdade de Ciências e Tecnologia, Universidade dos Açores, Campus de Ponta Delgada, Açores, Portugal

Abstract
Environmental non-governmental organizations (ENGOs) are groups of concerned citizens, working to improve the quality of life within communities, while preventing environmental degradation from economic activities. These groups assume an important role for populations inhabiting small islands, given their small territory, dependence on the import of goods and a focus on tourism to balance their fragile economies. In order to access the public commitment in developing sustainable tourism practices, key informant interviews were conducted with representatives of the ENGOs based on three archipelagos (Cabo Verde, Fernando de Noronha and Azores) aimed to characterize the ENGOs' mission and collect information on their projects and degree of involvement of local communities. An analysis of capital is applied to each research site in an effort to situate the role of ENGOs in their respective communities. The degree of commitment and official involvement in local tourism planning and management is different among the islands studied.

Keywords: island communities, environmental non-governmental organizations (ENGOs), community-based tourism, sustainable development

7.1 Introduction

Tourism is a valuable asset for any economy, but when it comes to communities living on islands, it is usually their largest source of income (Seetanah, 2011). Islands are small territories that encompass important natural values and growing populations, making it difficult to manage the trinomial of sustainability: economy, society and environment. The environment supports society, which in turn is responsible for the economy. Thus, only with a suitable and healthy environment can an economy thrive. Scott Cato (2009), in her book on green economics, argues that the economic pillar of sustainable development is a subsystem of society, which in turn is a subsystem of the environment, meaning that 'society should be embedded within the ecosystem, and markets and economies are social structures that should respond to social and environmental priorities' (publisher's description). The Worldwide Fund for Nature (2020) report acknowledges the ecological crisis humankind is facing, and islands, in particular, are seen as vulnerable to these threats (e.g. climate change, invasive species, biodiversity loss).

Islands with distinctive natural ecosystems often see their populations turn to tourism

*Corresponding author: maria.am.ventura@uac.pt

 Tourism Transformations in Protected Area Gateway Communities (eds S.L. Slocum *et al.*)
DOI: 10.1079/9781789249033.0007

as an income source. As such, islands have a long history of creating protected areas to act as valuable assets for habitat conservation, which often become hotspots for nature-based tourism (Queiroz *et al.*, 2014a; Kurniawan *et al.*, 2016). However, tourism can be responsible for cumulative impacts leading to environmental degradation. While the literature identifies multiple ways to tackle this problem (Queiroz *et al.*, 2014b), best practice necessitates the involvement of local communities to reduce environmental degradation (Jamal and Stronza, 2009; Bennett *et al.*, 2012).

Community involvement in local tourism management can be achieved through non-governmental organizations (NGOs) and community-based groups of citizens engaged in seeking solutions to environmental problems. NGOs are well known for their role in safeguarding the long-term interests of citizens, offering the public a route to dialogue with powerful stakeholders (Jeffrey, 2001). NGOs are also perceived to exercise a growing influence over business and act as regulatory agencies (Nalinakumari and MacLean, 2005). The environmental NGOs (ENGOs) are increasingly influential in conservation science, employing scientists whose studies support the fight against major biodiversity threats (Fonseca, 2003). While tourism is considered a developmental opportunity to insular regions (Fortes, 2016), from an ENGO perspective the tourism industry is far from being innocuous (United Nations Conference on Sustainable Development, 1999). Tourism may not be a sustainable development option in areas where environmental legislation does not exist or is not enforced (Gössling, 2000). ENGOs that operate within communities are helping to shape the theoretical foundations of sustainable tourism concepts (Ellis and Sheridan, 2014).

Estima (2017) analysed the role of ENGOs operating on islands and gathered evidence demonstrating the role of ENGOs in encouraging sustainable tourism. Her findings further emphasized the importance of involving ENGOs in the development of a sustainable tourism model, in a participative and inclusive manner. The archipelagos under scrutiny by Estima were the Azores, a Portuguese outermost region, and Fernando de Noronha (FN), an archipelago offshore of the Brazilian state of Pernambuco, for which the important role of ENGOs to the sustainability of the tourism industry was demonstrated (Estima *et al.*, 2013, 2014, 2015, 2016, 2017). Our work focuses on three Atlantic archipelagos, from the north to the south: the Azores (part of Portugal), the Republic of Cabo Verde and Fernando de Noronha (part of Brazil) (Fig. 7.1). All have an enormous potential for tourism due to their recognized natural and cultural heritage. These heritage values are recognized by the United Nations Educational, Scientific, and Cultural Organization (UNESCO), which has classified numerous elements in each archipelago as having 'exceptional world interest'. Each archipelago has several protected areas (PAs) with management plans, the effectiveness of which is often doubtful due to the lack of good surveillance systems or to scarcity of human and financial resources (Pieraccini *et al.*, 2017).

7.2 Literature Review

Islands are biodiversity hotspots, both inland and along the coast. The presence of legally protected fragile areas implies restrictions to their use. Legal protection infers the existence of management plans, but the lack of proper supervision usually prevents their full implementation (Baldwin and Black, 2008). According to the United Nations World Tourism Organization (1998), sustainable tourism development meets the needs of present tourists and host regions, while protecting and enhancing opportunities for the future. It is envisaged as leading to the management of resources so that economic, social and aesthetic needs can be fulfilled, while maintaining cultural integrity, essential ecological processes, biological diversity and life support systems. It also highlights that inclusive community development should be at the heart of tourism policies through education, investment, innovation and technology, which can transform livelihoods within island communities while preserving the environment and culture, and drive a more inclusive and sustainable form of tourism (United Nations World Tourism Organization, 2020).

Tourism can be particularly important for small island destinations. Given their challenges and limitations – specifically, isolation, limited economic opportunity and fragile resources – these

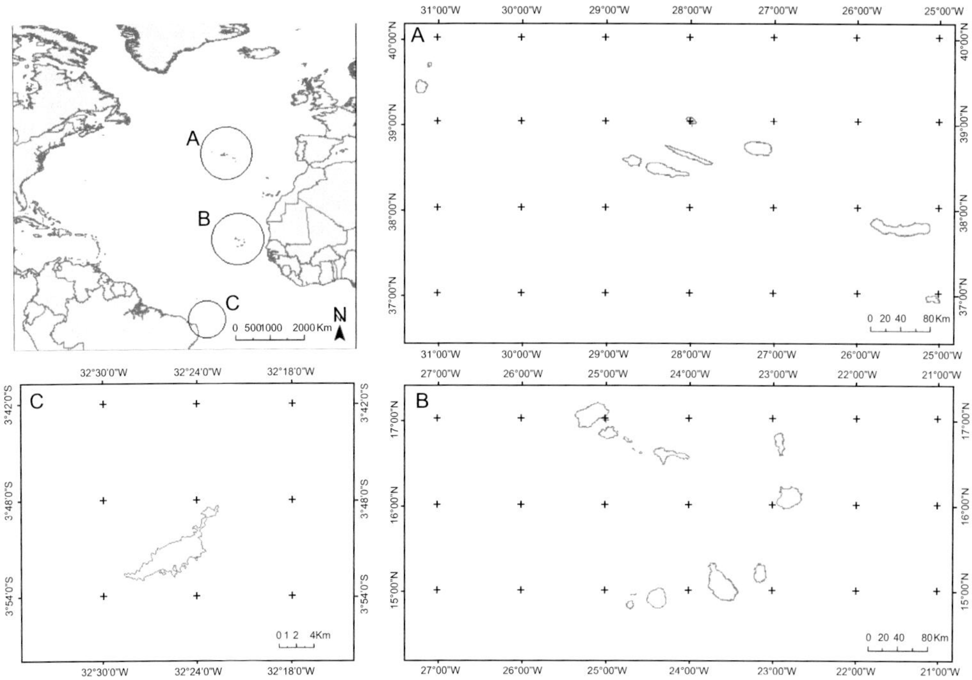

Fig. 7.1. Location of study sites: (A) Azores; (B) Cabo Verde; (C) Fernando de Noronha (figure courtesy of the authors).

destinations need to prioritize pro-sustainable planning and management, putting into practice all principles of sustainability (environmental, social, economic, political, territorial, cultural, symbolic, philosophical). By involving local actors at all stages, a more ethical, inclusive, responsible and participatory development model can be formulated (Estima, 2017).

Environmental problems gained importance as the degradation of habitats and biodiversity loss became the basis of a social movement in the 1970s and 1980s (Holden, 2016). Environmental consciousness and levels of public participation started to rise in countries with long and consolidated democratic histories (Wenqi, 2020). Specific to our research sites: Brazil ended its military dictatorship in 1985; Portugal regained its democracy in 1974; and Cabo Verde, which became independent from Portugal in 1975, experienced political openness from 1990 onward. These political changes have facilitated the appearance of ENGOs in these archipelagos. In Cabo Verde, the first NGO emerged in 1987, following the publication of Law 28/III/87 of 31 December that regulates the exercise of freedom of association (PlatONG, 1999). Although the Azores arrived later to this battle for 'civic responsibility', the Azoreans have been experiencing a growing environmental awareness, especially now that the archipelago has become the first in the world to be certified as a sustainable tourism destination according to the Global Sustainable Tourism Council (GSTC) standards (Earthcheck, 2019). This certification benefits the promotion of the destination, but brings with it responsibilities to meet the GSTC standards. Thus, it is an opportunity for local ENGOs to enhance pressure on their policy-makers.

ENGOs are powerful actors in the role of civil society (Nalinakumari and MacLean, 2005). The active involvement of ENGOs in the decision-making process of tourism management in PAs is assured by their presence as deliberative or consulting political bodies. In Cabo Verde, most NGOs and community associations are represented on the NGO Platform (Plataforma das

ONGs – PlatONG), created in 1996 to facilitate communication among their members (PlatONG, 2015). They are also present in the Conselho Nacional do Ambiente (Environmental National Council), a governmental body. The government resolution on the National Strategy of Protected Areas (Resolution No. 35/2016) states (article 2, point d) that NGOs must be allowed to participate in the processes of the creation and selection of PAs. Finally, and because many of these ENGOs develop work on the conservation of sea turtles, they are also members of the national network for the protection and conservation of marine turtles (Rede Nacional de Protecção e Conservação das Tartarugas Marinhas – TAOLA). This network involves not only the ENGOs, but also governmental and research bodies, such as the National Environmental Body (Direção Nacional do Ambiente – DNA), the Instituto Nacional de Desenvolvimento das Pescas, a fisheries research institute, and a few city councils that run conservation projects in Cabo Verde. In Fernando de Noronha, ENGOs are members of the Conselho Consultivo das Unidades de Conservação (Consulting Board of Conservation Units), a board that merges two previous organizations: the Environmental Protection Area and the Fernando de Noronha Marine National Park (Estima *et al.*, 2017). This board, linked to the management structure of the conservation units, is a forum to debate environmental, social and economic challenges regarding the management of these conservation units. In the Azores, ENGOs are members of the Conselho Regional do Ambiente e do Desenvolvimento Sustentável (Regional Board for the Environment and Sustainable Development), which acts as a consultant body for the government, and whose main mission is based on the Aarhus Convention (1998), ratified by Portugal in 2003.

7.3 Research Sites

The Atlantic Ocean offers beautiful and scenic groups of inhabited islands that rely on tourism as an important source of income. Islands of each archipelago have unique natural attractions that offer different experiences to visitors. On small islands, like the ones in our study, most people live in close proximity to PAs and support themselves by providing goods and services to tourists. They also directly experience the negative impacts of increased tourism visitation and are subject to certain restrictions put in place in the conservation areas. Thus, by definition, they can be defined as gateway communities (Bennett *et al.*, 2012).

7.3.1 Cabo Verde

Cabo Verde (Figs 7.2, 7.3, 7.4 and 7.5) consists of ten volcanic islands and several islets lying off the west coast of Africa with a population of around 563,000 people (Instituto Nacional de Estatística, 2018). The archipelago is naturally fragmented over 58,000 km^2 of ocean, isolated from permanent continental influences, maintains a relatively rich tropical biodiversity (Medina *et al.*, 2015) and is considered an important marine ecoregion (Spalding *et al.*, 2007) and nesting bird area. It has coralline communities, the second-largest nesting area of sea turtles (*Caretta caretta*) in the North Atlantic (third largest in the world) and is an important breeding ground for humpback whales (Wenzel *et al.*, 2009).

Given the economic and social potential of marine biodiversity, marine and coastal ecotourism (such as underwater sightseeing tours, whale watching and nesting sea turtles) is an emerging economic activitiy in Cabo Verde (Medina *et al.*, 2015). However, ecotourism may be impaired due to the disruption of marine ecosystems, if hunting pressures and destruction of habitats is not minimized (Taylor and Cozens, 2010; National Directorate for the Environment, 2015; Medina *et al.*, 2015). The 2010–2013 Strategic Plan for Tourism Development highlights the main natural spaces that make up the PA network of the islands and emphasizes the need for conservation and the enhancement of wildlife, flora and landscape values as key tourism products (National Directorate for the Environment, 2015). These values are also reiterated in the National Strategy for the Blue Economy (Cabo Verde Resolution No. 172/2020).

With biodiversity under strong anthropogenic pressure, mitigation measures have been implemented. Cabo Verde Decree-Law No. 3/2003 establishes the legal regime to manage coastal and

Fig. 7.2. Fogo Natural Park, Island of Fogo, Cabo Verde (photo courtesy of the authors).

Fig. 7.3. Natural Reserve (Reserva Natural) Boavista Island, Cabo Verde (photo courtesy of the authors).

Fig. 7.4. Natural Park of Cova/Ribeira da Torre/Paul, Island of Santo Antão, Cabo Verde (photo courtesy of the authors).

marine PAs according to the importance of their biodiversity, natural resources and ecological, socioeconomic and touristic interest. There are presently 46 PAs established on six islands, 21 of which are coastal or marine parks, representing 12.96% of the territory. In 2016, the National Protected Areas Strategy was approved, which established guidelines for a long-term expansion plan, control strategies for climate change and the provision for civil society, local communities and NGOs to participate in the creation and selection of PAs (National Directorate for the Environment, 2015). Moreover, Maio and Fogo islands were recognized as biosphere reserves by UNESCO in 2020 and, on Santiago island, the Historic Centre of Ribeira Grande in Cidade Velha was acknowledged as a UNESCO cultural heritage site in 2009.

7.3.2 Fernando de Noronha

Fernando de Noronha (Figs 7.6, 7.7, 7.8 and 7.9) comprises 21 islands and islets off the Brazilian coast, with a total area of 26 km^2, whereof 9 km^2 is the marine platform and 17 km^2 is exposed land. In Fernando de Noronha, there is no private land, only state areas, and the population of approximately 5000 lives in areas under the jurisdiction of the State District of Fernando de Noronha (Lopes *et al.*, 2017; Cristiano *et al.*, 2020). Fernando de Noronha archipelago is a UNESCO natural heritage site, along with Rocas Atoll Reserve, recognized for its unique marine biodiversity, geological processes and singular ecosystems. The archipelago represents a large portion of the South Atlantic island surface, and its waters are extremely important for the breeding and feeding of tuna, shark, turtle and marine mammals (Pires *et al.*, 2016). The Fernando de Noronha marine PA (MPA) is divided into two management categories: a no-take zone (controlled tourism allowed) and a sustainable use zone (dwelling, fisheries and tourism permitted). The Chico Mendes Institute for Biodiversity Conservation, an autonomous federal agency under the Ministry of the Environment, is responsible for the management and conservation of the site that

Fig. 7.5. ENGO BIOS.CV, Boavista Island, Cabo Verde (photo courtesy of the authors).

has two separate management plans, one for Fernando de Noronha and the other for Rocas Atoll. These plans address issues like tourism, research, environmental education, protection and biodiversity monitoring.

For the Fernando de Noronha Marine National Park, a Sustainable Development and Ecotourism Management Plan was implemented with local people's support (UNESCO, 2001). It controls the development of tourism infrastructures, visitation and the urbanized areas outside the park. This plan considers the carrying capacity of different regions within the park and regulates navigation, diving and management practices in this fragile environment (UNESCO, 2001). However, Feitosa and Gómez (2013) identified that tourism negatively impacts the island ecosystem, particularly due to fuel-burning from planes and the power generation needed to meet tourism demand. Moreover, spinner dolphins have been impacted by the intensification of touristic activities. Vegetation has also been altered by anthropogenic actions, particularly by wood extraction, fires, exotic species invasion and deforestation for agriculture (Pagano, 2000).

Fernando de Noronha Marine National Park has a concession agreement with EcoNoronha Company for the provision of services to support public visitation and ticket collection, as well as the sustainability and management of ecotourism (Estima *et al.*, 2014). It also attempts to strengthen the relationship between the PA and the gateway communities, including actions to improve the life quality of the local population, overcome conflicting relations in the community, and manage institutions and protect tourism interests (Lopes *et al.*, 2017).

Fig. 7.6. View of an area of PARNAMAR, the FN Marine Park, Fernando de Noronha Island, Brazil (photo courtesy of the authors)

Fig. 7.7. Baia do Sancho, beach inside the PARNAMAR area, Fernando de Noronha, Brazil (photo courtesy of the authors).

Fig. 7.8. Praia do Sueste, FN Marine Park, Fernando de Noronha, Brazil (photo courtesy of the authors).

7.3.3 The Azores

The Azores archipelago (Figs 7.10, 7.11, 7.12 and 7.13) comprises nine inhabited islands and some islets. Tourism began in the 19th century, when the Azores started receiving recognition in the travel literature (e.g. Henriques, 1867; Weeks, 1882). Since then, the naturalness of the islands has attracted many travellers, and tourism is perceived as a high strategic sector that can contribute to the development of the region (Vieira *et al.*, 2013; Ponte *et al.*, 2018). Fragmentation and degradation of habitats, together with the spread of non-indigenous species (Silva *et al.*, 2008; Elias *et al.*, 2016), are the greatest threats to Azorean biodiversity and are perceived to be promoted by increasing tourism activities (Parretti *et al.*, 2020; Costa *et al.*, 2021).

The relatively high endemism level provides conservation relevance to the Azores' biota. Its preservation has been recognized by the local government through the establishment of PAs for conservation since the early 1980s (Borges *et al.*, 2006). The regional network of PAs (Azores Regional Legislative Decree No. 15/2007/A) consists of nine island natural parks and one marine park (Calado *et al.*, 2011, 2012). The nine island natural parks include 23 sites of community importance and 15 special PAs. These PAs include the richest sites of endemic species; therefore, attention needs to be paid to the implementation of sustainable uses of the archipelago's natural resources, in a trade-off with human activities (Borges *et al.*, 2006).

The Azores Geopark combines protection and promotion of the Azores' geological heritage with a socioeconomic, educational, environmental and cultural focus. A strong emphasis is put on geotourism, a holistic approach focused on valuing the volcanic nature of the territory and its footprint on biodiversity, culture, architecture, traditions and ethnography (UNESCO, 2017). The islands of Corvo, Flores and Graciosa, and the coastal lagoons in São Jorge Island, are part of the Man and Biosphere Program that seeks to establish a scientific basis for enhancing the relationship between people and their environments, combining natural and social sciences to improve human livelihoods, while safeguarding natural and managed ecosystems (UNESCO, 2019).

Fig. 7.9. ENGO TAMAR in the Environmental Protection Area of Fernando de Noronha, Brazil (photo courtesy of the authors).

7.4 Methodology

This research encompasses ENGOs from three Atlantic archipelagos: Cabo Verde, the Azores (Portugal) and Fernando de Noronha (Brazil). The information was gathered by applying key informant interviews, according to the methodology by Kumar (1989), with descriptive statistics collected during the interviews. This primarily qualitative type of data collection method with key informants (Ali *et al.*, 2013) helped to identify ENGOs' missions and

Fig. 7.10. Corvo Island sighted from Flores Island, Azores (photo courtesy of the authors).

Fig. 7.11. Waterfalls at Fajã Grande, in the Natura 2000 area Costa Noroeste, in Flores Island, Azores (photo courtesy of the authors).

Fig. 7.12. Whale watching between the Islands of Pico and Faial, Azores (photo courtesy of the authors).

Fig. 7.13. Lagoa dos Cubres in São Jorge Island, Biosphere reserve, Azores (photo courtesy of the authors).

challenges, and gaps within their respective communities. Whenever possible, the head of each organization or a close representative was interviewed to guarantee a similar weight of expert opinions within the study. To compare our data with previous information, the semi-structured key informant questionnaire was partly based on Estima (2017). The face-to-face interviews were conducted with at least one representative of each Cape-Verdean ENGO during the year 2019. For the ENGO Associação Fauna e Flora de São Francisco, which changed leadership during the data collection process, the interview was conducted with both presidents.

At least one ENGO per island in the archipelago was approached, except for São Nicolau, which was omitted due to travel difficulties. The interviews were audio-recorded with permission. Additional information was gathered using participant observation and informal talks with volunteers from the ENGO. The ENGOs herein include: 18 in Cabo Verde (Table 7.1), 3 in Fernando de Noronha (Table 7.2), and 15 in the Azores (Table 7.3).

The interview consisted of two parts. Part 1 was meant to assess both the profiles of the responding person (mainly age and gender), and demographics of the ENGO (date of creation, scope and origin, and main mission); here we also explored the ENGO's main constraints on developing their actions, and the perceived degree of success on societal problem-solving. Part 2 had questions concerning the involvement of the ENGO in tourism and the existence of financed projects supporting the related activities. For each project mentioned, people interviewed were asked to give a degree of importance (from 1 – very important to 3 – not important) relating to five areas that they believed their projects would impact: 'contribute to public policies'; 'promote environmental conservation'; 'promote the island tourism sustainability'; 'promote the economic and social development of the community'; and 'promote the integration of local community and tourists'. Descriptive statistics were used to summarize the data. Fig. 7.14 provides an overview of the number of ENGOs in each research site and the number of tourism projects.

7.5 Results

7.5.1 ENGO characteristics – interview Part 1

Of the 18 ENGOs in the Cabo Verde archipelago, 90% were established after 1990. Concerning the origin of the ENGOs, Santiago and Boa Vista had five ENGOs each, while the remaining islands had either two (Santo Antão, São Vicente) or one (Sal, Maio, Fogo, Brava). Of the ENGO members interviewed, there was an even distribution of men and women, and the majority were young adults (66% under 40 years old). Regarding the ENGOs' missions in Cabo Verde, 'conservation of natural heritage' or particular related elements (e.g. sea turtles, sharks, terrestrial flora) were mentioned by all the ENGOs. Other goals, such as 'conflict management', 'community development', 'community training (e.g. young people, artisanal fisherman)' or 'development of ecotourism activities', appear to be increasingly important. The primary limitation faced by these agencies was a lack of funding, a concern shared by 70% of the ENGOs in Cabo Verde, as well as the inability to interact with politicians. It is worth noting that Cabo Verde ENGOs expressed some unique concerns not shared by the other research sites in this study, such as bureaucracy, difficulty in establishing partnerships with governmental bodies, the enforcement of legislation, lack of surveillance, insecurity to perform surveillance activities and management problems.

In Fernando de Noronha, all ENGOs appeared in the mid-1980s (although the Environmental Institute of Fernando de Noronha only formalized its existence in 2001). It has three ENGOs, two of which are local. Study participants were also equally distributed between men and women. In Fernando de Noronha, 'conservation and research' appear to be their primary focus. The absence of specialized personnel was the main limitation, as well as limited interactions with politicians.

For the Azores, one ENGO was established in 1963, and 66.7% appeared after the millennium, 50% of which materialized during or after 2010. The Azores also register a high percentage of

Table 7.1. Names of ENGOs in Cabo Verde, year of creation, scope of operation and main mission.

Country	ENGO	Creation date	Scope	Main mission
Cabo Verde	Associação Cruzinha em Santo Antão	1991 (officialized in 2003)	Regional (island of Santo Antão)	Environmental and social awareness, conservation of natural heritage
	Projeto Vitó – Santo Antão	2011	Regional (island of Santo Antão)	Conservation of sea turtles in the Porto Novo area (Santo Antão)
	Biosfera 1	2006 (officialized in 2010)	Regional (island of São Vicente/Windward Islands)	Protection and conservation of CV coastal and marine biodiversity
	Fundação Tartaruga	2012	National (origin in the island of Boa Vista)	Conservation of sea turtles and conflict management
	ADAD (Associação para a Defesa do Ambiente e Desenvolvimento)	1991	National (origin in the island of Santiago)	Protection of the marine and terrestrial environment
	Associação Fauna e Flora de São Francisco	2005 (officialized in 2011)	Regional (island of Santiago)	Protection of sea turtles and terrestrial flora
	Associação Varandinha de Povoação Velha	2009 (officialized in 2013)	Regional (island of Boa Vista)	Community development in the various sectors of economic activity
	Bi Flores – Brava	2016	Regional (island of Brava and its islets)	Conservation of ecosystems and protection of marine biodiversity (turtles and sharks), and terrestrial flora
	BIOS.CV	2012	Regional (island of Boa Vista)	Conservation of threatened species, their habitats, and promoting sustainable development
	Associação Ambiental Caretta caretta	2014	Regional (island of Santiago)	Conservation of sea turtles and wetland species
	ECO-CV (Associação Caboverdiana de Ecoturismo)	2015	National (origin in the island of Santiago)	Training of young people, collaboration in research on MPAs and conservation of marine species, development of an ecotourism network
	Associação Lantuna	2013	Regional (island of Santiago)	Conservation of biodiversity in natural areas and community training for the development of ecotourism activities
	MarAlliance	2015 in CV	Belize (In CV: islands of Boa Vista, Sal, Maio, Santiago)	Conservation of the marine environment, especially sharks and rays; training of artisanal fishermen

Continued

Table 7.1. Continued

Country	ENGO	Creation date	Scope	Main mission
	Cabo Verde Natura 2000	1998 (officialized in 2003)	Regional (island of Boa Vista)	Conservation of species and habitats, community development and preservation of the archaeological and architectural heritage
	ADRSCA (Associação Desportiva, Recreativa, Social, Cultural e Ambiental da Ponta do Pom)	1990 (environmental aspect in 2007)	Regional (Islands of São Vicente, Santo Antão)	Performance in sports, recreational, social, cultural aspects; in the environmental aspect, awareness of the conservation of sea turtles
	Projeto Biodiversidade do Sal	2015	Regional (island of Sal)	Protection of local biodiversity (sea turtles and birds) and creation of economic opportunities for the population
	Associação Projecto Vitó Fogo	2009 (officialized in 2010)	Regional (islands of Fogo (origin); Brava, Santiago, Boa Vista, Santo Antão)	Conservation of natural resources in the Fogo, Brava and islet regions, and ensuring the sustainable use of natural resources and their integration into tourism
	Fundação Maio Biodiversidade	2010	Regional (island of Maio)	Conservation of marine (turtles, cetaceans, birds, sharks) and terrestrial biodiversity; sustainable community development and promotion of ecotourism activities

Table 7.2. Names of ENGOs in Fernando de Noronha, year of creation, scope of operation and main mission.

Region (country)	ENGO	Creation date	Scope	Main mission
Fernando de Noronha (Brazil)	TAMAR (Pro-Tamar Foundation)	1984 in FN (1980 in Brazil)	National	Research and conservation of sea turtles
	IAFENO (Instituto Ambiental de Fernando de Noronha)	1988 (officialized in 2001)	Local	Conservation of the environment and participation in the creation of FN conservation units
	Golfinho Rotador	1990	Local	Spinner dolphin research and monitoring

Table 7.3. Names of ENGOs in the Azores, year of creation, scope of operation and main mission.

Region (country)	ENGO	Creation date	Scope	Main mission
Azores (Portugal)	Círculo de Amigos de São Lourenço	1988	Local (island of Santa Maria)	Promote, requalify and protect São Lourenço Bay
	CADEP (Clube dos Amigos e Defensores do Património Cultural e Natural de Santa Maria)	1991	Local (island of Santa Maria)	Valorization, safeguarding and dissemination of the cultural and natural heritage of Santa Maria (Branch of ENGO Amigos dos Açores)
	Associação Amigos do Mar de Santa Maria	2012	Local (island of Santa Maria)	Protection and promotion of the underwater natural heritage for scuba-diving
	OVGA (Observatório Vulcanológico e geotérmico dos Açores)	2000	Regional (location on the island of São Miguel)	Science center for research and scientific dissemination in the area of volcanology and geothermal
	Associação Ecológica Amigos do Calhau	2006	Local (island of São Miguel)	Defending nature, environment and coastal heritage
	Associação Ecológica Amigos dos Açores	1984 (officialized in 1989)	Local (island of São Miguel)	Defending and valuing the environment, as well as promoting nature conservation
	SPEA (Sociedade Portuguesa para o Estudo das Aves)	2003 in the Azores (1993 in mainland Portugal)	National	Conservation of birds and their habitats, from a sustainable point of view
	Associação Os Montanheiros	1963	Local (origin in Terceira island; other islands, São Miguel, São Jorge and Pico)	Defense of the natural heritage of the Azores, caving activities, training, tourism promotion (exploration of two show-caves)
	Associação para o estudo do ambiente insular	2002	Local (island of Terceira)	Manages the Azores Environment Observatory, dissemination of environmental projects
	Associação cultural, desportiva e recreativa da Graciosa	1993	Local (island of Graciosa)	Promotes sociocultural, sports and recreational activities for young people, valuing natural and cultural heritage
	Associação dos Amigos da Caldeira de Santo Cristo	2010	Local (island of São Jorge)	Sustainable and integrated development of Fajã de Santo Cristo
	Associação dos Amigos da Fajã dos Vimes	2005	Local (island of São Jorge)	Protection of the natural, historical and built heritage of Fajã dos Vimes

Continued

Table 7.3. Continued

Region (country)	ENGO	Creation date	Scope	Main mission
	Associação GeoParque Açores	2010	Regional	Geoconservation of the geological heritage of the Azores; management of the Capelinhos Volcano Interpretation Center on the island of Faial
	ASPEA (Associação Portuguesa de Educação Ambiental – Base Açores)	2015 in the Azores (1990 in mainland Portugal)	National	Environmental education in partnership with Turismo dos Açores and Observatório do Mar; organization of scientific tourism events
	Associação de defesa do ambiente da ilha das Flores	2012 (officialized in 2015)	Local (island of Flores)	Protection of the environment and the natural heritage of Flores, a UNESCO heritage island

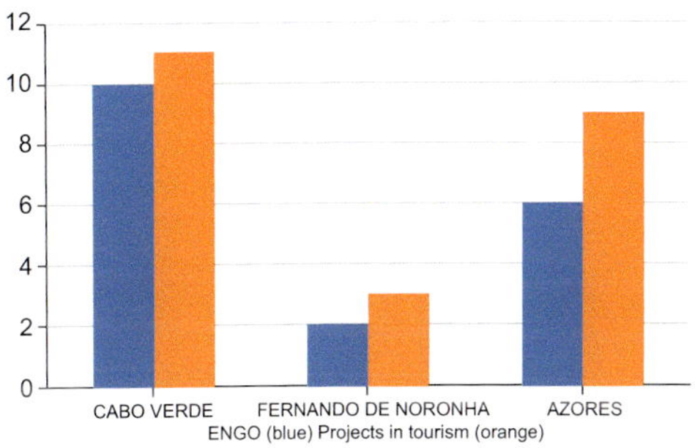

Fig. 7.14. ENGOs (blue) and tourism projects (orange) per study site (figure courtesy of the authors).

locally based ENGOs (67%). Participants included a majority of men (87%) above 41 years old (67%). The primary goals were 'cultural/natural heritage', which was reported by 60% of the ENGOs, 'promote, protect, and defend' as mentioned by 47%, and 'environmental education and sustainable development dissemination' stated by 33% of the participants. The Azorean ENGOs claim a shortage of funds as their primary limitation (except for the oldest, well established ENGO).

7.5.2 ENGO involvement in tourism – interview Part 2

Comparing the number of ENGOs involved in projects linked to tourism within each archipelago, Fernando de Noronha presents the highest percentage (67%) and the Azores the lowest (40%) (Fig. 7.15). When asked about the degree of importance on the perceived impact of their projects (based on five proposed sentences), we obtained the results presented in Fig. 7.16. ENGOs in Cabo Verde placed more value on 'promote environmental conservation' and 'promote the integration of local community and tourists' and valued less 'contribute to public policies'. The Azores also valued 'promote environmental conservation' the highest, followed by 'contribute to public policies' and valued least 'promote the integration of local community and tourists'. Because Fernando de Noronha has only two ENGOs involved in tourism projects, 'promote the economic and social development of the community' is the only statement that received maximum values. 'Promote the integration of local community and tourists' was the least valued statement.

7.6 Discussion

The discussion section situates the study results within the overarching policies of each research site and within the literature in an attempt to position the role of ENGOs, their mission, and challenges as they relate to site-specific political environments. The goal is to add context to the results.

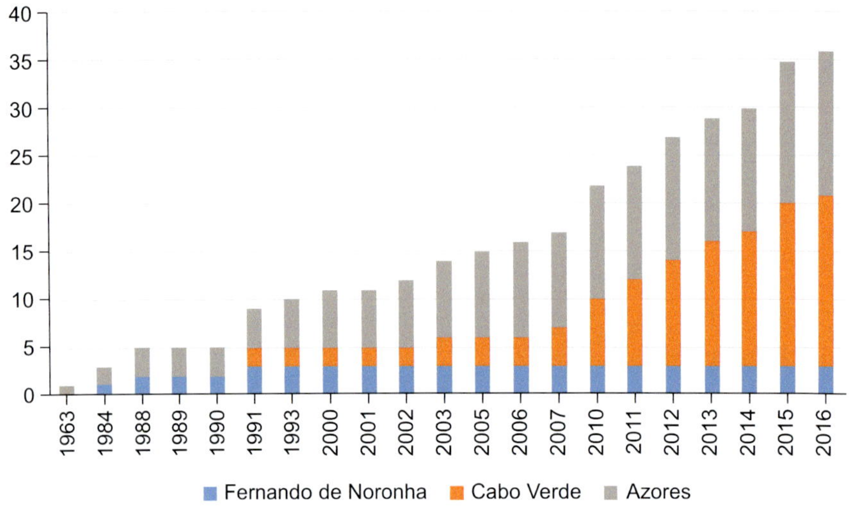

Fig. 7.15. Tourism related projects along time (FN – Fernando de Noronha; CV – Cabo Verde; AZ – Azores) (figure courtesy of the authors).

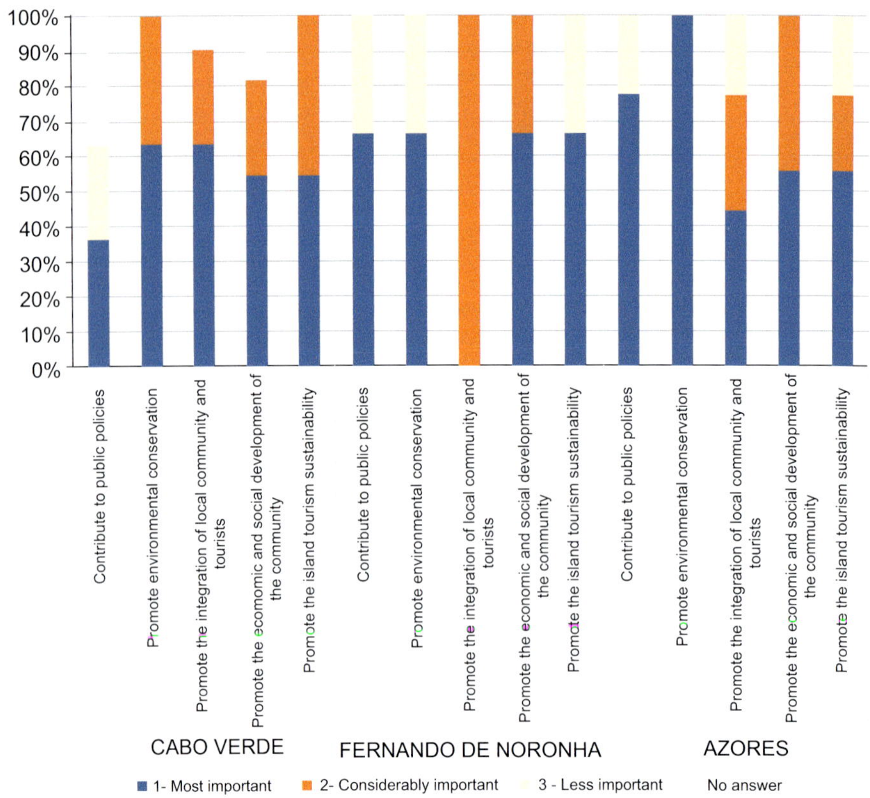

Fig. 7.16. Prioritization of agency goals on tourism projects (figure courtesy of the authors).

Calado *et al.* (2012) stress that in marine spatial planning, NGOs can be both stakeholders and facilitators that help to solve conflicts, an important mission identified by some of the Cabo Verde ENGOs. Community development and training were also important goals to the Cabo Verde ENGOs, acknowledging their role in the development of a more inclusive society. To the ENGOs of all three archipelagos, the environmental element (conservation and sustainable development) is a top priority. Ghina (2003) firmly asserts that a healthy environment is the basis of all life support systems, and that environmental vulnerability is a hallmark of these small Atlantic islands. Thus, the concerns expressed by local ENGOs recognize conservation and sustainability as primary objectives of their organizations.

Whale and bird watching are two important tourist segments in the Azores (Bentz *et al.*, 2016), however only SPEA (Sociedade Portuguesa para o Estudo das Aves), a national ENGO, has a mission devoted to protecting the Azorean avifauna. The remaining Azorean ENGOs have a wider focus on natural heritage. Different scenarios are found in the missions of ENGOs from the other two tropical archipelagos. Cabo Verde and Fernando de Noronha are both important sea turtle breeding areas, whose species are all considered vulnerable or endangered by the International Union for the Conservation of Nature (2020). One of the three existing ENGOs in Fernando de Noronha is especially devoted to the conservation of sea turtles (Tamar Foundation), another to the conservation of spinner dolphins (Golfinho Rotador). In Cabo Verde, only 4 of the 18 ENGOs studied did not mention sea turtles as an element of concern. Moreover, one ENGO, imported to Cabo Verde from Belize (MarAlliance), prioritizes shark and ray conservation. Other key biodiversity elements, such as birds, terrestrial flora or cetaceans, are also mentioned in the mission of most of these ENGOs. These natural elements can be a major attraction to visitors, so by promoting conservation actions for sea turtles and their nesting areas, or shark populations, these ENGOs boost ecotourism and nature-based tourism, two growing segments of island tourism worldwide (e.g. Donohoe and Needham, 2008; Slinger-Friedman, 2009). In Cabo Verde, three ENGOs (ECO-Cabo Verde, Associação Lantuna, Fundação Maio Biodiversidade) include training for ecotourism activities in their mission. Moreover, on the islands of Sal and BoaVista, which are experiencing greater tourist development (Banco de Cabo Verde, 2018), a well-known international tourist operator finances ENGOs' conservation projects aimed at protecting marine turtles and promoting environmental education actions. This is an implicit recognition of the importance of biodiversity elements to act as 'flagship species' which attract tourists.

In Fernando de Noronha, the absence of specialized personnel was the main limitation. This forces the ENGOs to import people from mainland Brazil, thus increasing the already high population, a limitation also perceived by four Cabo Verde ENGOs. The Azorean ENGOs claim a shortage of funds, except for the oldest, well established one. Important financing partners for Cabo Verde are represented by governmental bodies, municipalities and international organizations (e.g. United States Fish and Wildlife Services; MAVA Foundation (Switzerland); Global Environment Facility of the United Nations; Darwin Initiative (UK)).

7.6.1 Assessment of access to capital

Using the definitional framework for capital assets given by Bennett *et al.* (2012), we consider that Fernando de Noronha and the Azores have consolidated political and natural capital, while Cabo Verde is still in the process of improving its political capital and defining the value of its natural capital. Physical and built capital in Fernando de Noronha is private, supervised by the Chico Mendes Institute for Biodiversity Conservation. In the other two archipelagos, besides private initiatives always subjected to existing legislation, the federal government maintains responsibility. Human capital is perceived as sufficient in the Azores, but insufficient in Fernando de Noronha and Cabo Verde, particularly on the smaller islands. Some of the Fernando de Noronha and Cabo Verde ENGOs perceive the need to import human capital as threatening to the environmental sustainability of the islands. Cultural capital is equally important to the three archipelagos, particularly in the Azores and Cabo Verde, where it is considered vital to improve tourists' experiences. Social capital is, perhaps, what best connects these insular Atlantic territories. Islanders are people that, given their isolation from the nearest continents, tend to turn towards themselves to engage in join efforts to overcome adversities posed by the islands (Rodríguez-Díaz, 2018).

Financial capital differs between the archipelagos according to their financing opportunities. The Azores, as an outermost region of the European Union (EU), benefits from EU initiatives like the Observatory on Tourism for Islands Economy, an initiative that promotes cooperation and partnerships between EU insular territories, with NGOs offering data and research to support strategic planning documents (Haase and Maier, 2021). In Fernando de Noronha, the tax revenue paid by tourists is meant for island conservation so that it continues to meet tourists' needs for a satisfactory experience. Cabo Verde is the archipelago that has more financial constraints (expressed by local ENGOs), but being an emerging economy and a safe African country to visit puts it in a privileged position to receive financing from United Nations programmes and other important partners, such as aid funding from the USA or EU. The EU is willing to cooperate with Cabo-Verdean society through the civil society organizations, which include many international NGOs (Topliyski *et al.*, 2020).

In 2019, the authors of the study visited Fernando de Noronha and had the opportunity to speak with volunteers from the local ENGO. It was discovered that apart from the limitations already felt, there were some new concerns mainly related to the high cost of living in Fernando de Noronha. This is occurring due to the high price of the Environmental Preservation Tax (Taxa de Preservação Ambiental) paid by tourists to enter the island, plus the fee paid to EcoNoronha to access the Fernando de Noronha Marine Park. These taxes have impacted the price of goods and services for local residents on the archipelago. Another major concern was the uncertainty of the future of sustainability on the island, since in the last three years the residents felt that the carrying capacity of the island (number of tourists) had already been surpassed. While not the main focus of this chapter, these issues seem to create additional challenges within gateway communities.

7.7 Conclusion

On small islands, urban and natural areas coexist with communities living in or near the surroundings of PAs. In these Atlantic archipelagos, their gateway communities depend greatly on tourism. A partnership model of community-based ENGOs (Jamal and Stronza, 2009) can help to adopt more friendly solutions for tourism development. The role of these organized groups of local citizens in the development of a sustainable tourism model has been witnessed in the three archipelagos: Cabo Verde – submitting the first proposal of MPAs for the islands of Sal, Boavista and Maio to value nature conservation and ecotourism; Fernando de Noronha – empowering the local community to develop sustainable tourism actions; and the Azores – bringing to the state's attention the importance of creating PAs for marine tourism (e.g. scuba-diving). ENGOs are not only part of these island gateway communities but give them an opportunity to participate actively in local tourism management.

This study is limited to the perceptions of ENGO administrators. Future research should connect these results with the perceptions and opinions of gateway community members to ensure that local communities are actively engaged in the work of ENGOs. Moreover, ENGOs are primarily focused on conservation efforts. How these organizations balance the needs of residents with those of conservation could provide a more nuanced understanding of ENGO involvement in gateway communities.

Acknowledgements

This work is part of a H2020-MSCA-RISE-2016 project: *Planning in a liquid world with tropical stakes: solutions from an EU-Africa-Brazil perspective – PADDLE*. Funding has also come from FCT – Fundação para a Ciência e a Tecnologia, and the European Union, QREN, FEDER, COMPETE, by funding the CIBIO/InBIO (project UID/BIA/50027/2020 and POCI-01-0145-FEDER-006821) and postdoc grant (ref.M3.1.a/F/025/2016) by Fundo Regional de Ciência e Tecnologia (FRCT), AÇORES 2020. We would like to thank all representatives of the ENGO involved in the research, and to Cabo-Verdean political and research bodies for providing data useful to our research. Thanks are also due to Deborah Estima for filing some gaps on our data, and two anonymous reviewers that helped to improve the manuscript.

References

Aarhus Convention (1998) *Convention on Access to Information, Public Participation in Decision-making and Access to Justice in Environmental Matters*. Aarhus, Denmark, 25 June 1998.

Ali, M., David, M. and Ching, L. (2013) Using the key informants interview (KIIs) technique: a social sciences study with Malaysian and Pakistani respondents. *Man and Society* 24, 131–148.

Azores Regional Legislative Decree No. 15/2007/A. Diário da República n.° 120/2007, Série I de 25 de Junho.

Baldwin, R. and Black, J. (2008) Really responsive regulation. *The Modern Law Review* 71(1), 59–94.

Banco de Cabo Verde (BCV) (2018) *Relatório do estado da economia de Cabo Verde em 2017*. Departamento de Estudos Económicos e Estatísticas, Ed. Available at: http://www.bcv.cv (accessed 4 February 2022).

Bennett, N., Lemelin, R., Koster, R. and Budke, I. (2012) A capital assets framework for appraising and building capacity for tourism development in aboriginal protected area gateway communities. *Tourism Management* 33(4), 752–766.

Bentz, J., Lopes, F., Calado, H. and Dearden, P. (2016) Enhancing satisfaction and sustainable management: whale watching in the Azores. *Tourism Management* 54, 465–476.

Borges, P., Amorim, I., Gabriel, R., Cunha, R., Frias Martins, A. *et al.* (2006) Azores. In: Gillespie, R. and Clagu, D. (eds) *Encyclopedia of Islands*. University of California Press, Berkeley, CA, USA, pp. 70–75.

Cabo Verde Decree-Law No. 3/2003 Conselho de Ministros. Boletim oficial da República de Cabo Verde No. 5, I série, 24 Fevereiro.

Cabo Verde Resolution No. 172/2020 Conselho de Ministros. Boletim oficial da República de Cabo Verde, No. 138, I Série, 21 Dezembro.

Cabo Verde Resolution No. 35/2016 Conselho de Ministros. Boletim oficial da República de Cabo Verde, No.17, I Série, 17 Março.

Calado, H., Ng, K., Lopes, C. and Paramio, L. (2011) Introducing a legal management instrument for offshore marine protected areas in the Azores Marine Park. *Environmental Science and Policy* 14(8), 1175–1187.

Calado, H., Bentz, J., Ng, K., Zivian, A., Schaefer, N. *et al.* (2012) NGO involvement in marine spatial planning: a way forward? *Marine Policy* 36, 382–388.

Costa, A., Balibrea, A., Raposeiro, P., Santos, S., Souto, M. *et al.* (2021) Non-Indigenous and invasive freshwater species on the Atlantic islands of the Azores archipelago. *Frontiers in Ecology and Evolution* 9, 1–19.

Cristiano, S., Rockett, G., Portz, L. and Souza Filho, J. (2020) Beach landscape management as a sustainable tourism resource in Fernando de Noronha Island (Brazil). *Marine Pollution Bulletin* 150, 110621. DOI: 10.1016/j.marpolbul.2019.110621.

Donohoe, H. and Needham, R. (2008) Internet-based ecotourism marketing: evaluating Canadian sensitivity to ecotourism tenets. *Journal of Ecotourism* 7(1), 15–43.

Earthcheck (2019) *The Azores - the World's First Certified Archipelago*. Available at: https://earthcheck.org/news/2019/december/the-azores-the-worlds-first-certified-archipelago/ (accessed 31 October 2020).

Elias, R., Gil, A., Silva, L., Fernández-Palacios, J., Azevedo, E. *et al.* (2016) Natural zonal vegetation of the Azores Islands: characterization and potential distribution. *Phytocoenologia* 46(2), 107–123.

Ellis, S. and Sheridan, L. (2014) A critical reflection on the role of stakeholders in sustainable tourism development in least-developed countries. *Tourism Planning & Development* 11(4), 467–471.

Estima, D. (2017) *O papel das ONGA na sustentabilidade do turismo em ambientes insulares: estudo de caso Brasil – Portugal*. Doctoral dissertation. Universidade de Aveiro, Portugal.

Estima, D., Martins, M., Rabinovici, A. and Ventura, M. (2013) A atuação das Organizações Não-Governamentais ambientalistas no turismo em ilhas: o caso de Fernando de Noronha (PE, Brasil). *Revista Brasileira de Ecoturismo* 6(1), 153–170.

Estima, D., Ventura, M., Rabinovici, A. and Martins, F. (2014) Concession in tourism services and partnerships in the Marine National Park of Fernando de Noronha, Brazil. *Revista de Gestão Costeira Integrada* 14(2), 215–232.

Estima, D., Ventura, M., Rabinovici, A. and Martins, F. (2015) As ONGA e o turismo costeiro e marinho: contribuições para uma gestão participativa na ilha de Santa Maria, Açores. Actas do VIII Congresso sobre planeamento e gestão das zonas costeiras dos países de expressão portuguesa, 17.

Estima, D., Ventura, M., Rabinovici, A. and Martins, F. (2016) Gestão de usos e conflitos em áreas marinhas: o caso da ilha de santa maria, portugal. *Anais do II Congresso Ibero-Americano de Gestão Integrada de Áreas Litorais* 81–84.

Estima, D., Ventura, M.A., Rabinovici, A. and Martins, F. (2017) Condicionantes e contributos das organizações não governamentais de Aabiente para a sustentabilidade do turismo em ilhas: Estudo de caso do arquipélago dos Açores. *Revista Turismo & Desenvolvimento* 27/28, 1627–1637.

Feitosa, M. and Gómez, C. (2013) Aplicação do tourism ecological footprint method para avaliação dos impactos ambientais do turismo em ilhas: um estudo em Fernando de Noronha. *Revista Brasileira de Pesquisa Em Turismo* 7, 220–238. DOI: 10.7784/rbtur.v7i2.509.

Fonseca, G. (2003) Conservation science and NGOs. *Conservation Biology* 17(2), 345–347.

Fortes, C. (2016) Avaliação ambiental em Cabo Verde. Master Thesis. Faculdade de Ciências da Universidade do Porto, Portugal.

Ghina, F. (2003) Sustainable development in small island developing states. *Environment, Development and Sustainability* 5, 139–165.

Gössling, S. (2000) Tourism – sustainable development option? *Environmental Conservation* 27(3), 223–224.

Haase, D. and Maier, A. (2021) *Research for REGI Committee – Islands of the European Union: State of Play and Future Challenges*. European Parliament, Policy Department for Structural and Cohesion Policies, Brussels, Belgium.

Henriques, B. (1867) *A trip to the Azores or Western Islands*. Lee and Shepard, Boston, MA, USA.

Holden, A. (2016) *Environment and Tourism*. Routledge, London, UK.

Instituto Nacional de Estatística (2018) *Estatísticas do Ambiente 2016*. Instituto Nacional de Estatística de Cabo Verde. Available at: https://ine.cv/publicacoes/estatisticas-do-ambiente-2016/ (accessed 4 February 2022).

International Union for the Conservation of Nature (2020) *Marine Turtle Red List Assessments* SSC Marine Turtle specialist group. Available at: https://www.iucn-mtsg.org/statuses (accessed 4 February 2022).

Jamal, T. and Stronza, A. (2009) Collaboration theory and tourism practice in protected areas: stakeholders, structuring and sustainability. *Journal of Sustainable Tourism* 17(2), 169–189. DOI: 10.1080/09669580802495741.

Jeffrey, D. (2001) The roles of environmental non-governmental organisations in the twenty-first century. *Biology and Environment: Proceedings of the Royal Irish Academy* 101B(1-2), 151–156.

Kumar, K. (1989) *Conducting key informant interviews in developing countries – A.I.D. Program design and evaluation methodology report n° 13*. United States Agency for International Development.

Kurniawan, F., Adrianto, L., Bengen, D. and Prasetyo, L. (2016) Vulnerability assessment of small islands to tourism: the case of the Marine Tourism Park of the Gili Matra Islands, Indonesia. *Global Ecology and Conservation* 6, 308–326.

Lopes, P., Mendes, L., Fonseca, V. and Villasante, S. (2017) Tourism as a driver of conflicts and changes in fisheries value chains in Marine Protected Areas. *Journal of Environmental Management* 200, 123–134. DOI: 10.1016/j.jenvman.2017.05.080.

Medina, A., Gomes, I., Araujo, S., Lima, L. and Monteiro, R. (2015) *Fifth National Report on the Status of Biodiversity in Cabo Verde*. Available at: https://www.cbd.int/kb/record/nr/105205?Country=cv (accessed 4 February 2022).

Nalinakumari, B. and MacLean, R. (2005) NGOs: A primer on the evolution of the organizations that are setting the next generation of regulations. *Environmental Quality Management* 14(4), 1–21.

National Directorate for the Environment (2015) *Fifth National Report on the Status of Biodiversity in Cabo Verde*. Available at: https://www.cbd.int/doc/world/cv/cv-nr-05-en.pdf (accessed 4 February 2022).

Pagano, S. (2000) *Crescimento Descontrolado ou Desenvolvimento Sustentável: a encruzilhada do turismo em Fernando de Noronha*. Master Thesis. Universidade Federal de Pernambuco, Brasil.

Parretti, P., Canning-Clode, J., Ferrario, J., Marchini, A., Botelho, A. *et al*. (2020) Free rides to diving sites: the risk of marine non-Indigenous species dispersal. *Ocean Coastal Management* 190, 105158.

Pieraccini, M., Coppa, S. and De Lucia, G.A. (2017) Beyond marine paper parks? Regulation theory to assess and address environmental non-compliance. *Aquatic Conservation: Marine and Freshwater Ecosystems* 27(1), 177–196.

Pires, N., Garla, R. and Carvalho, A. (2016) The economic role of sharks in a major ecotourism archipelago in the western South Atlantic. *Marine Policy* 72, 31–39.

PlatONG (1999) *Enquadramento Jurídico das ONG's em Cabo Verde*, Fevereiro de 1999. Praia. Available at: https://issuu.com/plataformadasongspong/docs/enquadramento_juridica_das_ong (accessed 6 April 2021).

PlatONG (2015) *Guia das ONGs de Cabo Verde*. Plataforma das ONGs de Cabo Verde, 254 pp. Available at: http://www.laboress-afrique.org/ressources/assets/doc/Document_N0109.pdf (accessed 6 April 2021).

Ponte, J., Couto, G., Pimentel, P. and Oliveira, A. (2018) Tourism activities and companies in a sustainable adventure tourism destination: the Azores. *Tourism & Management Studies* 14(4), 25–38.

Queiroz, R., Ventura, M. and Silva, L. (2014a) Plant diversity in hiking trails crossing Natura 2000 areas in the Azores: implications for tourism and nature conservation. *Biodivers Conservation* 23, 1347–1365.

Queiroz, R., Ventura, M., Guerreiro, J. and Cunha, R. (2014b) Carrying capacity of hiking trails in Natura 2000 sites: a case study from North Atlantic Islands (Azores, Portugal). *Journal of Integrated Coastal Zone Management* 14(2), 233–242.

Rodríguez-Díaz, C.E. (2018) Maria in Puerto Rico: Natural Disaster in a Colonial Archipelago. *AJPH HURRICANES* 108(1), 30–32.

Scott Cato, M. (2009) *Green Economics: An Introduction to Theory, Policy and Practice*. Routledge, London, UK.

Seetanah, B. (2011) Assessing the dynamic economic impact of tourism for island economies. *Annals of Tourism Research* 38(1), 291–308.

Silva, L., Ojeda Land, E., Rodríguez Luengo, J., Borges, P., Oliveira, P. *et al*. (2008) *Invasive terrestrial flora and fauna of Macaronesia. TOP 100 in Azores, Madeira and Canaries*. ARENA, Ponta Delgada, Azores.

Slinger-Friedman, V. (2009) Ecotourism in Dominica: studying the potential for economic development, environmental protection and cultural conservation. *Island Studies Journal* 4(1), 3–24. DOI: 10.24043/isj.225.

Spalding, M., Fox, H., Allen, G., Davidson, N., Ferdaña, Z. *et al*. (2007) Marine ecoregions of the world: a bioregionalization of coastal and shelf areas. *Bioscience* 57, 573–583.

Taylor, H. and Cozens, J. (2010) The effects of tourism, beachfront development and increased light pollution on nesting Loggerhead turtles *Caretta caretta* (linnaeus, 1758) on Sal, Cape Verde Islands. *Zoologia Caboverdiana* 1(2), 100–111.

Topliyski, A., Labrunie, I., Cruz, D., Sousa, N. and Tavares, E. (2020) *Mapeamento da Sociedade Civil Cabo-verdiana*. União Europeia - Assistência técnica e Capacitação das Organizações da Sociedade Civil e Autoridades Locais. Available at: http://www.platongs.org.cv/wp-content/uploads/2021/03/Mapeamento-Sociedade-Civil-CV.pdf (accessed 4 February 2022).

United Nations Conference on Sustainable Development (1999) *Sustainable Tourism: A Non-governmental Organization Perspective*. United Nations Conference on Sustainable Development, Department of Economic and Social Affairs, Commission on Sustainable Development, Background Paper, p. 4.

United Nations Educational, Scientific, and Cultural Organization (2001) *Brazilian Atlantic Islands: Fernando de Noronha and Atol das Rocas Reserves*. World Heritage Centre. Available at: https://whc.unesco.org/en/list/1000 (accessed 25 October 2020).

United Nations Educational, Scientific, and Cultural Organization (2017) *Açores UNESCO Global Geopark (Portugal)*. Available at: http://www.unesco.org/new/en/natural-sciences/environment/earth-sciences/unesco-global-geoparks/list-of-unesco-global-geoparks/portugal/acores/ (accessed 25 October 2020).

United Nations Educational, Scientific, and Cultural Organization (2019) *Man and the Biosphere (MAB) Programme*. Available at: https://en.unesco.org/mab (accessed 25 October 2020).

Vieira, J., Couto, G., Pimentel, P., Menezes, A., Moniz, A. *et al*. (2013) The satisfaction of the Nordic Tourist with the Azores as a destination. *Scandinavian Journal of Hospitality and Tourism* 13(1), 59–73.

Weeks, L. (1882) *Among the Azores*. James Boston, Osgood & Company, Boston, MA, USA.

Wenqi, D. (2020) How culture shapes environmental public participation: case studies of China, the Netherlands, and Italy. *Journal of Chinese Governance* 5(3), 390–412.

Wenzel, F., Allen, J., Berrow, S., Hazevoet, C., Jann, B, *et al*. (2009) Current knowledge on the distribution and relative abundance of humpback whales (*Megaptera novaeangliae*) off the Cape Verde Islands, Eastern North Atlantic. *Aquatic Mammals* 35, 502–510. DOI: 10.1578/AM.35.4.2009.502.

World Tourism Organization (WTO) (1998) *Guide for Local Authorities on Developing Sustainable Tourism*. United Nations World Tourism Organization, Madrid, Spain.

World Tourism Organization (WTO) (2020) *AlUla Framework for Inclusive Community Development through Tourism*. United Nations World Tourism Organization, Madrid, Spain.

Worldwide Fund for Nature and Living Planet Report (2020) *Bending the Curve of Biodiversity Loss*. World Wildlife Fund, Gland, Switzerland.

8 Protected Area Tourism in Batanes Islands, Philippines: Issues and Impacts

Edieser Dela Santa[1]*, Mary Anne Ramos-Tumanan[1] and Chris Ryan[2]
[1]University of the Philippines, Philippines; [2]University of Waikato, New Zealand

Abstract
Various frameworks have been employed in the monitoring and evaluation of tourism impacts in protected areas (PAs). Aware of their limitations, this chapter adopts a longitudinal, stakeholder-driven approach to examine the emerging impacts to gateway communities near PAs in Batanes, the Philippines, as part of a larger initiative to establish a sustainable tourism observatory in the islands. Findings show perception of impacts to be interconnected with discourses on the role of tourism in development, conflicts over resource use and place identity, and reveal the strength of a discursive approach. Anchored in stakeholder engagement, eliciting factors that underlie tourism impacts in a PA include weak local governments with strong central administrations focused on maximizing economic growth over other concerns, such as sustainability, and robust place attachment driven by rootedness. They point to the importance of improving governance and raising the possibility of co-management as a mechanism to reform PA management.

Keywords: Batanes Islands, stakeholder engagement, protected areas (PAs), tourism development

8.1 Introduction

Various methods have been employed in the monitoring and evaluation of tourism impacts in protected areas (PAs). To investigate visitor impacts, indicator-based frameworks, such as visitor impact management (VIM), visitor experience and resource protection (VERP), the recreation opportunity spectrum (ROS), and similar approaches, have been proposed. These frameworks have been applied in a range of contexts including marine PAs (e.g., Farrell and Marion, 2002; Fefer *et al.*, 2018). However, many frameworks take a linear, one-shot or one-off thematic approach to the study of tourism impacts. The result is oftentimes a short-term evaluation of the effects of tourism, focused on narrow aspects, such as economic, social or environmental. This potentially magnifies or diminishes the causative value of tourism because these approaches lack a holistic perspective.

Aware of the limitations of these approaches, this chapter adopts a longitudinal and stakeholder-driven approach to examine the emerging impacts of PA tourism. It utilizes a Foucaultian notion of discourse that Foucault perceived as being fundamental to societal facts and conditions. The starting point of the Foucaultian analysis is structured by the development of diverse and unified concepts (Vandaele, 2016), constructed (or deconstructed) by the conditions of their actual existence. A Foucaultian approach seeks to uncover

*Corresponding author: eddelasanta1@up.edu.ph

DOI: 10.1079/9781789249033.0008

what are seen as 'changing points of weakness' (Ife, 2013, p. 61) pertaining to issues and problems that arise as a consequence of structured and unstructured thoughts. This chapter argues that this strategy is able to identify the complex and interdependent nature of social, cultural, economic and environmental factors that underpin tourism impacts. The approach, germane to the processes of research centres under the auspices of the United Nations World Tourism Organization International Network of Sustainable Tourism Observatories (UNWTO-INSTO), is consistent with transdisciplinary methods to assess PA tourism.

The research takes the case of Batanes, the only province declared by Philippine law as having protected landscape and seascape status, based on the strength of its outstanding natural resources and the cultural practices and heritage of the Indigenous Ivatans community. Currently, the islands are on the tentative list of the United Nations Educational, Scientific and Cultural Organization's (UNESCO) World Heritage Sites and efforts are underway to have the province included in the UNWTO-INSTO. In terms of gateway communities, the islands, from one perspective, are places through which visitors arrive to destinations, while in return they are, for their own residents, the gateway to other places. The Batanes islands are generally a recipient destination into which tourists are funnelled by plane and boat. Equally, the Batanes islands are the gateway to one of the world's richest marine areas, as characterized by the wealth of the pink and red corals found there, which are among the world's rarest. These resources have propelled tourism growth, with tourist arrivals breaching 50,000 prior to the pandemic, three times that of the local population.

Through discourse analysis, the chapter shows that this growth is very much related to sustainability issues, namely economic, environmental and sociocultural, all of which are interconnected and bound within the discussion on, and construct of, Batanes as a PA. This is carried out by first describing Foucaultian discourse analysis, place attachment, the significance of UNWTO-INSTO, and the co-management issues, particularly assessments, that face gateway communities to PAs. After explaining the methodology, the situation in Batanes is then described. It is emphasized that the range of interrelated issues identified by the chapter shows the strength of a discursive approach, anchored on stakeholder engagement, in eliciting factors that underlie tourism impacts in a PA gateway community.

8.2 Review of Related Literature

8.2.1 Understanding Foucaultian discourse analysis

Foucault notes a complementarity between a power–knowledge nexus and a Marxian theory of capitalism (Diaz-Bone *et al.*, 2007). In the modern age, he suggests that a great deal of power and social control is exercised not by 'brute physical force or even by economic coercion, but by the activities of experts who are licensed to define, describe, and classify things and people' (Cameron, 2001, p. 16). This means that the language in use ('practice', as termed by Foucault) is a vehicle to control and dominate ideas, people and methods of behaviour. He purports that a discourse is not pure content or a spoken language per se, but acquires meaning with certain characteristics that necessitate reflection and questioning in order to understand the context. He alludes that 'the individual subjects are not only its inert target, they are the elements of its articulation. The individual, which power has constituted, is at the same time its vehicle (Foucault, 1980, p. 98).

As noted by Ife (2013), Foucaultian discourse analysis (FDA) infers that production of knowledge, actual or real, is grounded in circumstances that require a move beyond a structured perspective to better understand how issues are defined, and how they occur and recur through discursive discourses. His view complements an academic tradition that Foucault calls a 'discourse' whose currency is under the guise of new ways of thinking, of musing. While Foucault used the term 'discourse' in different ways (Akdağ and Swanson, 2018, p. 71), FDA, as a methodological approach, examines the recurrent patterns through themes peripheral to impacts, in this case through tourism development, interjecting positions and consequences that may be present within pressing issues. Therefore, FDA brings attention to new ways of looking at the context and layers of intertwined issues. Foucault defines discourses as 'practices which systematically form the objects of which they speak' (Foucault, 1972; cited in Cameron, 2001), and other scholars have thought afresh about practices and agencies involved throughout these exercises (e.g. Diaz-Bone *et al.*, 2007).

Recently, Ramos-Tumanan and Ryan (2019) examined the impacts of tourism in Chinese World Heritage Sites (Hongcun, Xidi, and Nanping) against the backdrop of place attachment and place meaning discourses. Their study illuminates the methodological consequences of marrying two research traditions by combining a methodological approach with Foucault's discourse analysis. They suggest that researchers need to adopt a gestalt perspective in seeking to analyse the underlying context within a discourse, to examine what is and what is not. They add that:

> Any discourse is as much articulated by what is not said, or remains silent, as much as by what is said. Indeed, often the silent is not recognized because it is assumed to be a common, an assumption, not worthy of examination because it is shared. However, often the sharing is hidden from sight, and then when examined, is found not to be homogenous.
>
> (Ramos-Tumanan and Ryan, 2019, p. 290)

Several critical aspects can be deduced from their study. First, it reinforces a reflective process to find hitherto silent nuances to fully understand the cultural and political contexts of tourism-induced change. Second, the examination of methodological consequences at the onset of the study requires familiarity with the research site, and, equally significant, an understanding that tourism research is context-bound. Thinking empirically, a good dataset does not always equate to having a good-fit model because data are grounded in experience. By observing how participants engage in daily activity, and in the process, the researcher can recognize the 'changing points of weakness' (Ife, 2013. p. 61) and the issues (occurring and recurring) under specific conditions, as they exist. This is where the reflective process manifests, as the researcher participates in the methodological process in which the wealth of a discourse rests upon its meaning dependent on the context, and not merely based on theoretical abstractions or on rules of hierarchical modelling. Arguably, 'the ability to engage with cultures different from those of the researcher is also an advantage when seeking to assess the impacts of tourism' (Ife, 2013, p. 291).

Some critiques argue that FDA is 'somewhat vague, with no clearly defined methodological structure for applying Foucault's concept' (Akdağ and Swanson, 2018), but Foucault himself (1974), noted that his methods were available for interpretation as the researcher deemed appropriate (cited in Akdağ and Swanson, 2018, p. 71). While FDA offers fruitful ways to understand the social constructions of meanings, the authors of this study are interested in identifying the recurrent challenges imposed by tourism and under which conditions these challenges occur.

8.2.2 Seeing outside-inside: place attachment–meaning discourses

> *Kung ano ang tanaw ay siyang nakikita;*
> *Kung ano ang nakikita ay siyang nararamdaman;*
> *Kung anu ang nararamdaman ay siyang ginagawa.*
> Translation:
> How you look is what you see;
> What you see is how you feel;
> How you feel is what you do.

In the above passage, Filipino Professor Emeritus F. Landa Jocano (1930–2013) suggests that the context of discourse may be associated with visuality, unconscious–conscious bias, emotions, space, place and broad tendency, to name a few. Scholars, such as Fyfe and Law (1988), Hollinshead (1999) and Herman (2016) have their own perception of the Filipino anthropologist's passage, viewed explicitly or otherwise, and distinctive of an individual or characteristic as basis of actions (Jocano, 1997). For instance, Fyfe and Law (1988, p. 1) note that 'image and visuality is a depiction, never just an illustration. It is the site for the construction and depiction of social difference', while Hollinshead, in his assessment of Urry's gaze (cited in Ramos-Tumanan and Ryan, 2019), affords the idea of a 'multi-ocular seeing of places ... gazes are also selective according to priorities attributed to what is thought important' (p. 290).

In her framework of place attachment and place meaning, Ramos-Tumanan (2018) develops a model suggestive of place attachment–meaning discourses. She suggests that place attachment is evidenced by attitudinal measures, namely cognitive, affective and conative. Tourism induces change, and how people respond to such changes goes beyond simple job and income generation. Initial studies were based within a western culture, and more recent work has indicated the importance of culture and the efficacy of local populations in planning as key components that determine successful outcomes (Ryan and Huang, 2013). Another area of the literature prioritizes the role of ethnic minorities and Indigenous groups to sustain such cultures (Ryan and Aicken, 2005).

In a seemingly complex discourse on the social process of place, Castells (2004) deduced that place meaning is constructed through a conflictive process triggered by the interests and values of opposing social actors. By extension, place identification is defined by a mutual set of interaction between the self and the larger sociocultural, economic, political and environmental conditions and structures in which the self is embedded (Wang and Xu, 2014). That concept has been increasingly tested and reinterpreted in the light of different cultures (Ramos-Tumanan and Ryan, 2019).

It follows, then, that the creation of sense of place corresponds to image creation, and the derivation of image today is made more complex by the increased use of social media. The literature is full of such models, and none have established predominance over others despite generalized theories, such as flow theory or gap theory. Some commentators have suggested that meta-narratives, based on generalization, may not be wholly possible for a number of reasons (Ryan *et al.*, 2011). Foremost within these is the view that tourism experiences are heavily contextualized, both temporarily and spatially, and that what is pertinent in one location may not apply to another. Another factor is the role of cultural understanding of experience, as different communities, while sharing many norms, attribute differing degrees of importance to various values (Ramos-Tumanan and Ryan, 2019).

8.2.3 The methodological significance of UNWTO-INSTO

The International Network of Sustainable Tourism Observatories (INSTO) is a monitoring body committed to a programme of research that goes beyond a single period of data collection. It is suggested that the initial stages of the proposed INSTO project be based on residents' perceptions of the impacts of tourism, as such information can be readily used by tourism planning and marketing bodies and would also permit an exchange of information with other INSTO network observatories (Ryan *et al.*, 2019). The INSTO programme is based on the United Nations Sustainable Development Goals (SDGs), with the ultimate goal of supporting better informed sustainable tourism policies. INSTO aims to achieve the following objectives, as indicated on its webpage (UNWTO-INSTO, 2019):

- develop and provide methodological references and recommendations for sustainability in tourism;
- create awareness and support capacity building in sustainability among tourism stakeholders while stimulating community participation;
- analyse and report regularly on policy trends and impacts on tourism;
- identify and disseminate good practices;
- facilitate exchange and comparison of information related to tourism sustainability;
- strengthen institutional capacities for information management and monitoring in support of decision and policy-making; and
- facilitate the establishment of a network of observatories to permit a global exchange of information.

The United Nations World Tourism Organization (UNWTO) set up the Statistical Framework for Measuring the Sustainability of Tourism, which has been incorporated into the INSTO framework to facilitate information exchange at various levels of governance, with measurements adapted to both local and national scales (Gúčik and Marciš, 2020). In terms of implementation, results have been encouraging, although large gaps remain. In an evaluation of perceived ease of use and usefulness of datasets, Niavis *et al.* (2019) found that INSTO indicators obtained low marks for sensitivity to data and stakeholder engagement. In comparison with other tourism sustainability assessment initiatives, managers rated the indicators highly for applicability and ease of use. Nonetheless, Niavis *et al.* caution that the scores were highly affected by how the individual projects perceived the construct 'sustainable tourism', underscoring the observation that a clear, needs-based conceptualization of the term is warranted to enable a robust sustainability assessment system.

As a peripheral location, and as the gateway to marine resources that attract UNESCO recognition, the Batanes experiences stressful conditions – on the one hand there is a need for economic resources that will benefit local Ivatans and the Filipinos at large, while on the other, sustainability and conservation are important. The implication for management would, therefore, seem to rest in audits of resources that can aid planning to remain within limits of acceptable change (Farrell and Marion, 2002). Such assessments would be part of a growing awareness among researchers regarding the need to investigate issues and challenges facing gateway communities to PAs (e.g., Joyner *et al.*, 2019; Stoker *et al.*, 2021).

8.3 The Case of Batanes Islands

Batanes is the northernmost province of the Philippines. The smallest province in the country, it comprises six municipalities and is located about 162 km north of the Luzon mainland in the Luzon Strait. Of this island group, the largest and only inhabited islands are Itbayat, Batan and Sabtang. It is home to the Ivatans, the province's Indigenous people. Batanes has a total population of 16,600, with 4270 households, as of 2018 (Philippines Statistics Authority, 2018). In 2018, it received more than 50,000 visitors, more than three times its resident population, and tourist arrivals are expected to increase in the coming years. Come *et al.* (2016) note that Batanes is one of the last remaining areas in the Philippines retaining unique natural physiographic features, including wave-cut cliffs, cave-like outcrops and secluded white sand beaches. This is attributed to its geographic location at the meeting point of the Pacific Ocean and South China Sea. Due to its geographical position, Batanes is home to distinct flora and fauna. Pink and red corals (*Corallum* sp.) abound in its marine environment. Of the 42 known flowering plants in the country, seven are found only in Batanes. Culturally, the province is the only area in the Philippines where traditional architecture is of stone (Fig. 8.1), in response to the winds and monsoon weather.

Because of these qualities, the Batanes group of islands was legislated in 1994 as an area of Protected Landscapes and Seascapes, the only province in the Philippines that has earned this distinction. Authorities are currently working to have it included in the UNESCO Intangible Cultural Heritage List. The monitoring and evaluation of tourism impacts on the island is not well established, like many other destinations in the country. To date, there has been no systematic assessment of the capacity of the islands, and their people, to support tourism and hospitality growth. Thus, the decision to set up a tourism monitoring centre, with aspirations to join the UNWTO-INSTO, was made.

8.4 Methodology

Co-management requires commonality of goals and modes of achieving those goals. Within international forums, such as that of the UNWTO, management processes will often have identified the above questions at an initial stage of any project. However, within national borders dominated by strong central governments, patterns of, often, unquestioned assumptions may inhibit true co-management. The 'authorities' can claim expertise, technical and moral, argue for the consistency of programmes across a country, and effectively, if unintentionally, conduct policies of colonialization within their own territories by seeking to disregard the local. Thus, one question for the Batanes team was to what extent was there a divorce between the representatives of the centre at the local level, and those who felt their legitimacy emerged from being local?

A range of research initiatives supported by grants from the University of the Philippines

Fig. 8.1. Stone houses are the main attraction in Chavayan, Sabtang (photo courtesy of Mary Anne Ramos-Tumanan).

were pursued between February 2019 and February 2020. The activities in each initiative aimed at stimulating the engagement of national and local government units and the acceptance of academe as partners, while at the same time promoting dialogue among key stakeholders. It started with the research team's effort to explore, in a broader sense, the overall situation in Batanes through several visits for observation, casual conversation with the locals and document reviews. This evolved into a more structured approach to collecting input on relevant tourism development and putting scenario-based issues to stakeholders at different levels and varying capacities. These scenarios related to policy, strategy and current practice. This method is consistent with the approach espoused by the UNWTO-INSTO programme and by scholars who value the discussion of issues associated with uncertainty (Capistrano and Notorio, 2020).

The results were restructured through the identification of common themes that, in turn, prompted priority issues for discussion. Chief among these activities were workshops and focus group discussions (FGDs) in July 2019 and January 2020 involving the public and private sectors, members of civil society groups and tribal elders from the three islands of Batan, Sabtang and Itbayat. A technical working group was commissioned to lead, but not dictate, the FGD in identifying priority areas in tourism and hospitality that will serve as the main research focus for the succeeding years. Inputs were also obtained from discussions with the Sangguniang Panlalawigan (provincial board) and relevant national government agencies overseeing PA tourism in Batanes.

8.5 Results

8.5.1 The interconnection of impacts from tourism growth

Similar to experiences observed in gateway communities (e.g., Joyner *et al.*, 2019), analysis

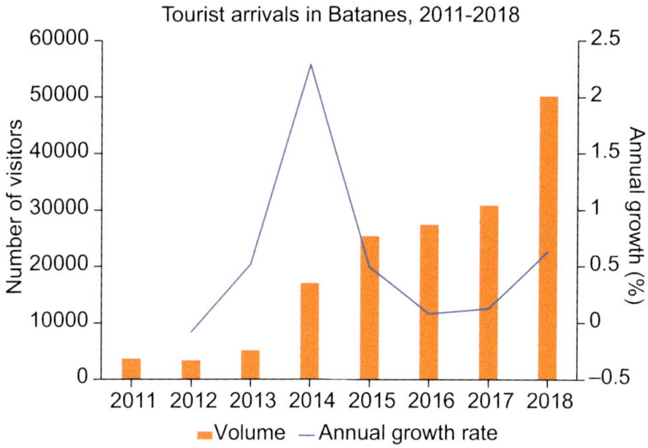

Fig. 8.2. Tourist arrivals in Batanes (figure courtesy of the authors).

of FGD data and secondary documents from Batanes reveals simultaneous issues relevant to food production and consumption that cut across other concerns, such as labour and employment, while a disastrous impact on solid and waste-water required immediate attention to land use planning side by side with spatial analysis. Additionally, differences between environment preservation and Indigenous people's rights weighed heavily, not least when Indigenous people's control over processes and structures was strained by top-down policy-making.

Participants noted that tourism was expanding rapidly in the province. Official statistics support this. In 2013, there were only 5000 visitors. Arrivals more than tripled the following year and, by 2018, visitor arrivals had exceeded 50,000 (Fig. 8.2) according to the Provincial Government of Batanes (2019). To meet this demand, there was an associated increase in accommodation establishments, restaurants, transportation services and tour operations. As a respondent stated, 'a lot of people set up homestays, hotels, and restaurants in the last five years'. The growth is, likewise, reflected in the number tourism-related organizations, such as the Batanes Homestay Association, Batanes Indigenous Cultural Community Organization, Sabtang Weavers' Association, and Batanes Heritage House Restoration and Conservation Program. The transport sector has also grown, with every municipality having a tricycle operator and driver's association (Fig. 8.3).

This expansion did not come by accident, rather, it is part of a plan that reflects the discourse on tourism as a panacea for long-standing societal issues, such as poverty alleviation and economic growth (Dela Santa and Saporsantos, 2016). Cemented as a national policy by the Tourism Act of 2009, which declared tourism as 'an engine of socioeconomic growth and cultural affirmation to generate investment, foreign exchange, and employment', the discussion is manifested in regional and local plans. Hence, according to the Cagayan Valley Regional Development Plan 2017-2022 (2019), Batanes, and the region as a whole, would be a source of water, an agro-industrial centre and a budding tourism destination. Under its own Batanes Ecotourism Master Plan, published in 2018, Batanes shall pursue a 'high value/low volume' strategy for economic prosperity, social development, environmental conservation and institutional development (Provincial Government of Batanes, 2018, p. 14).

The rapid development of tourism has exposed a number of sustainability issues, namely economic, environmental and sociocultural. These issues are interconnected, complex and all bound to the discourse on, and construct of, Batanes as a PA. From an economic perspective, stakeholders mention issues of food production and consumption – specifically, how

Fig. 8.3. Locally-designed tricycles to carry tourists around Batanes (photo courtesy of Edieser Dela Santa).

the supply of food (meat and fish in particular) ebbs and flows with tourism. One respondent said, 'We could not buy pork last year because they went straight to restaurants' (Fig. 8.4), and another explained, 'Locals could not buy quality fish, even if we went to the pier, as they had been reserved by hotels and restaurants'. The concern extends to the local delicacy *tatus*, the island's coconut crab (*Birgus latro*), which is now a threatened species.

This situation is connected to the inability of the islands to supply their own food needs. During the workshops, an official from the Department of Agriculture estimated that food sufficiency in the islands is at around 43%, which means that around two tons of food are imported by air every day, aside from supplies that arrive by cargo ships at the Port of Basco. Food insecurities spill over to related challenges, such as livelihood support and innovations for sustainable and productive agriculture and fisheries, which cut across food safety and quality. Local production is mostly focused on *wakay* (sweet potato), garlic, onion, ginger, *ube* (purple yam), *gabi* (tarot) and those crops less prone to destruction from typhoons. However, the nature of agricultural work relative to tourism employment has attracted workers to the latter, contributing to the downturn in agricultural production. As a respondent explained, 'in tourism, if I worked half a day as a tour guide, I'd get paid in the afternoon. In agriculture, I'd have to wait for a few months. And what if a typhoon came?'

This phenomenon is also reflected in the higher education sector of Batanes. As revealed in the workshops, tourism and hospitality courses are now the most popular among students of the state university in the province, while fishing and agriculture courses struggle to attract students. The result is high labour costs and challenges in attracting Ivatans to other occupations. The labour shortage has consequently led to the recruitment of workers from outside Batanes, according to a respondent, a phenomenon that residents had not seen prior to the expansion of tourism activities. The influx of visitors and transient workers has had the related effect of increasing demand for housing, with resulting higher property prices. It has been noted that the majority of accommodation facilities for tourism were formerly boarding houses for students and workers from the neighbouring Batanes islands. As explained by a respondent:

Fig. 8.4. Pork is hard to purchase during the peak season for tourism (photo courtesy of Mary Anne Ramos-Tumanan).

'Basco is where the two local colleges are found, so students from Sabtang and Itbayat come to rent. But because of the vibrancy of tourism, private houses here in Basco were turned into homestay facilities for tourists, instead of boarding houses for students. Students could not find anywhere to live anymore.'

Price inflation has been observed too: 'Land is sold and bought at 3000 to 5000 pesos per square metre now, which is too much for those [locals] seeking to establish residential houses.' Workshop participants also reported the conversion of agricultural lands into properties for tourism and residential purposes. All these, plus the spread of illegal (that is, not in the proper zone or not operating as permitted, such as a farmhouse) accommodation facilities owned by the Ipula (non-native of Batanes) raise the issue of improper planning that undermines the Ivatan's sense of place.

The economic pressure is exacerbated by the perceived environmental impacts of tourism. Residents have associated the increase in tourism activities with the surge in the volume of garbage, and a reduction of the water supply during peak times. As a respondent stated, 'The waste-water of homestays go to the canals, to creeks, then seep into the sea.' Of solid waste, a resident pointed out: 'There are no recycling centres, so plastic bottles and single use plastics keep on piling up. This is most felt, the garbage problem, in Basco because tourists stay here.' In short, numerous examples indicate that, as a gateway to marine resources, the islands are tipping beyond the levels of acceptable change, which can only have adverse effects on social and economic life.

Unfortunately, environmental expertise is also apparently in short supply. According to respondents, local authorities have had to rely on technical assistance from the Department of Environment and Natural Resources Environmental Management Bureau in Manila to resolve waste and water issues, resulting in a proposal to build a water treatment facility in Basco. This effort is meant, in part, to address concerns regarding the tourist carrying capacity of Batanes, which has not been assessed.

Stakeholders noted that there have been previous attempts to examine the issue, the last during the preparation of the Batanes Ecotourism Master Plan. Environment authorities admit to having a bias towards a quantitative measure that mainly considers density, but they have been stymied by data concerns.

8.5.2 Conflict over resource use

The importance placed on this highly objective approach extends to the domains of land tenure, resource management and cultural preservation within Batanes. As an official explains, in the case of cutting trees for use in heritage houses: 'If a lot is not titled, tree cutting is not allowed. We also need documentation of traditional practices involving extraction or harvesting of natural resources, in order to properly assess requests for exemption from the Department's rules.'

These rules include Republic Act No. 7586 that created the National Integrated Protected Areas System (NIPAS) Law and Proclamation No. 335, s. 1994, which declared Batanes islands as protected landscapes and seascapes. At the broad level, researchers (e.g., Ong, 2002) have noted how NIPAS Law conflicts with the Local Government Code of 1992, the Mining Act of 1995, the Indigenous Peoples Rights Act (IPRA) of 1997 and the Agriculture and Modernization Act of 1997. This is exacerbated by a lack of a national land use policy that would have shown boundaries between PAs and other land uses.

Stakeholders in Batanes, particularly culture and heritage advocates, lament the firm implementation of environmental laws, particularly because their actions are guided by the motivation, not just to preserve heritage, but also to empower the Ivatans under the IPRA Law. Questions, such as 'What cultural practices can be exempted from the NIPAS Act? What law takes precedence, NIPAS or IPRA?' remain and are a source of tension among stakeholders. These views are consistent with academic analyses of resource use and management in other PAs of the country (e.g. De Leon and Kim, 2017). Thus, to add to the problems of being a gateway to marine riches, there exists a lack of clarity over the legal status of cultural practices, where local communities have lived within the constraints imposed by natural resources of land and sea for generations.

8.5.3 Batanes as a PA, Batanes as a home

Amidst all these questions, Ivatans have found themselves revisiting the wisdom of being a PA, including whether it has benefited the province at all. A respondent summarized a sentiment commonly heard: 'The entire province is protected, which makes it hard to move. We cannot preserve our heritage, such as traditional houses, because our laws are strict. Without lifting certain parts, it will be very difficult.'

The latter point can be connected to the attachment Ivatans hold for Batanes as a place. It was pointed out by FGD participants that Batanes was a home first, before it was a PA, and accordingly, has been described in superlative terms. Said one resident, 'Batanes is a beautiful and peaceful place, the last place touched by God.' It is where 'harmonious interaction between men and nature is still observed,' added another. It is a 'safe haven', according to one respondent, who remembers a time when 'houses were only closed during typhoons'. Such sentiments and memories led another resident to proclaim that Batanes is where he is going to spend the rest of his life.

While these images evoke nostalgia, Batanes as a PA does not figure among interpretations mentioned by the Ivatans, revealing the top-down nature of the designation made in the 1990s under the recommendation of the Department of Environment and Natural Resources. As a respondent remembered, 'We were surprised by the announcement,' as apparently the general population was largely unaware of central administration initiatives to conserve priority PAs that include Batanes. Thus, while Ivatans purport to respect nature, the reverence was due to the islands being their home, and not because of following an administrative fiat imposed from above.

8.6 Discussion and Conclusion

The issues described above reflect a few points. First, the range of interconnected issues shows the strength of a discursive approach, anchored on stakeholder engagement, in eliciting factors that underlie tourism impacts in a PA. While approaches such as limits of acceptable change and VIM also include stakeholder inputs (Farrell and Marion, 2002), the power of a discursive approach comes from the identification of recurring ideas, regularities and connections (Joyner *et al.*, 2019). Through this method, the case of Batanes was able to show that the impacts of tourism can be wide-ranging. The emerging impacts are, likewise, complex, connected as they are to long-standing discourses on the role of tourism in economic growth, particularly in developing countries (Telfer and Sharpley, 2015), conflicts among Filipinos over the use of resources for tourism development (Dela Santa and Saporsantos, 2016; De Leon and Kim, 2017), and place-identity and development (Ramos-Tumanan and Ryan, 2019). Much of the current debate has seemingly missed the concept of the islands as a gateway to the marine environment, which, with the climate and weather, controls much of life's original patterns. In a U-turn from the conventional literature, where gateways are seen as points of control, via mechanisms such as access fees and land use plans (Joyner *et al.*, 2019), in the case of the Batanes, any development must note how nature, specifically typhoons, imposes constraints on the way in which tourism will be permitted to expand. Like many communities, those in the Batanes actually enjoyed the restrictions imposed by Covid-19 in much of 2020 and 2021. The islanders found higher levels of self-sufficiency in terms of food, less pollution, less environmental damage from tourist passenger movements, and more time for social interaction (personal communication with Batin and Velones, 26 January 2021).

The implications for methodological approaches, such as limits of acceptable change and VIM towards monitoring and evaluation of the impacts of PA tourism are clear: a stakeholder-driven approach is the key to 'seeing the unseen' and 'knowing the unknown' because the narratives identify the context of the pressing issues at different times and with different actors. After all, according to Niavis *et al.* (2019), tourism sustainability is broad enough to allow for different interpretations. In seeking to examine the unseeable and the unsayable aspects of a certain phenomenon implicit to the conditions and nuances within which it exists (Herman, 2016; Ramos-Tumanan and Ryan, 2019), different discourses on methodological strategies can be explored longitudinally, through engagement in transdisciplinary approaches that consider the multiplicity of voices in PA tourism planning. This is important, particularly for Batanes as a PA, since it has been observed by researchers (e.g., Dela Santa and Tiatco, 2019) that the prevailing narratives of some of Batanes' intangible culture are being contested by local communities because of alleged misrepresentations by the government and the tourism industry. Thus, they call for the plurality of voices, which ultimately identify ways in which tourism becomes a positive force for social change. This approach requires that discourse, as a source of evidence, needs methods that mean the researcher is 'more likely to spend time talking in depth with the participants, encouraging them to explore the subject in their own way and in their own words' (Cameron, 2001, p. 14).

The discourses on the interconnection of impacts from tourism growth, conflict over resource use, and place identity and tourism development, themselves, raise the importance of improving governance quality, which is broadly characterized by weak local governments, in a national setting. Governance is a thorny issue in the country. In the context of the Philippines, a number of observers have noted how destination performance is affected by institutional factors and broad structural conditions. Henderson (2011) explains that these include corruption, red tape and inadequate planning and programme implementation. Dela Santa (2015) directed attention to the personalistic nature of the system and asserted that unless the oppressiveness of these factors is mitigated, the context of Philippine tourism and hospitality policy and planning, including in relation to PAs, will remain very challenging.

This is where co-management might come in as a governance mechanism (Strickland-Munro *et al.*, 2010). While there is broad policy and institutional support for the concept at the national and the Batanes community level, as seen in the existence of a Protected Area Management Board where government and non-state actors discuss PA tourism, the discourses revealed above show that issues have not been resolved effectively at these forums. Investigating the reasons why is beyond the scope of this chapter, but some researchers (Weeks *et al.*, 2010; Nyaupane *et al.*, 2020) have observed how targets for PAs, including administrative effectiveness, have been difficult to fulfil and have, consequently, recommended reforming management systems. Perhaps a co-management approach tailored to the local context, which provides stakeholder access to technical expertise and facilitation to understand PA problems better (Strickland-Munro *et al.*, 2010), aided by systematic monitoring of indicators through mechanisms such as INSTO, could be explored. Through this, it might be possible for Batanes to minimize, if not prevent, the challenges and development pressures (Stoker *et al.*, 2021) captured in the discourses presented above and observed in other gateway communities.

Consistent with this approach is the in-depth investigation of the pluralist views and discourses identified by the chapter. Using FDA as a methodological lens, determinants of impacts have been identified that informed the consequences of tourism development in Batanes prior to the pandemic. Here, one can see areas for future work investigating two vital elements of Foucaltian power relations: the 'agents' who exercise power and the 'targets' who are the subordinate actors (Cheong and Miller, 2000). While the central government might easily qualify as the dominant authority in the discourse, the brokers from the private and public sectors, particularly local governments, have been given the autonomy to develop tourism facilities and attractions. What is the extent to which they participate in the perpetuation of the prevailing epistemological structure and logic? And what of tourists to Batanes? Cheong and Miller (2000) considered tourists as targets within the Foucaltian framework, as much of their activities are responsible for the propagation of principal views.

The above discussion represents the academic implications of this ongoing case study. It highlights a range of tasks and outputs from the stakeholders' workshops. Of note are evident concerns about the impacts of the rapidity of change on the Ivatans. Equally important are the implications for management. The findings reinforce the significance of engaging with gateway communities and stakeholders. It was apparent during the workshops that some stakeholders hold strong, yet conflicting, views related to the PA status of the islands, the inflow of migrant workers and the lack of food supply during the peak tourist season. It is important that these issues are addressed. While there are mechanisms for interaction, it is clear that such co-management schemes are proving to be inadequate in addressing community concerns. Informal and formal discussions through alternative channels, such as tourism-related organizations, could broaden engagement and increase understanding of issues.

The findings, likewise, stress the importance of setting up a more robust tourism and tourism-related data collection system than currently exists. Such systems are at the heart of sustainable tourism assessment models (Niavis *et al.*, 2019). In the context of the Philippines, this is challenging at the lower administrative level, where the balance of political ecology shows local governments struggling to keep up with the requirements of laws passed at the national level (Maguigad, 2013). However, for PAs such as Batanes, that are experiencing rapid tourism growth and suffering the consequences thereof, such a monitoring system, with well-developed indicators, is paramount to understand the relationship between tourism impacts and the ecological and cultural base on which sustainable tourism strategies depend (Kristjánsdóttir *et al.*, 2018).

The need to work together raises the associated issue of improving the capacity of institutions involved in the tourism and PA governance system (Nyaupane *et al.*, 2020). In Batanes, the consultations showed that only one of six municipalities had a permanent tourism officer; the rest had temporary staff. Moreover, the individuals concerned held other responsibilities, such as social work, human resources` and business licensing, thus severely limiting their time for tourism management.

Capacity building is easier said than done, particularly in gateway communities to PAs, given manpower and resource constraints, as seen in the Batanes islands. Yet, taking advantage of recent opportunities for collaboration, a hand-holding approach under a broader multi-stakeholder management framework (Strickland-Munro *et al.*, 2010) could be implemented as a co-management scheme. With stakeholders effectively engaged, driven, in no small part, by strong place attachment, the complexity of assessment and the interconnectedness of tourism impacts will be less difficult to understand. In turn this could facilitate the implementation of modalities for PA tourism towards conservation, sustainability and resilience.

Acknowledgment

The authors acknowledge the Office of the Chancellor of the University of the Philippines Diliman, through the Office of the Vice Chancellor for Research and Development, for funding support through Project No. 191909 ORG and Project No. 191905 PhDIA. They also thank the Provincial Government of Batanes, Council of Elders and other stakeholders from the province.

References

Akdağ, E. and Swanson, D. (2018) Ethics, power, internationalisation and the postcolonial: a Foucauldian discourse analysis of policy documents in two Scottish universities. *European Journal of Higher Education* 8(1), 67–82.

Cagayan Valley Regional Development Plan (CRVDP) 2017-2022 (2019) Tuguegarao City, Philippines: National Economic Development Authority Regional Office 2. Available at: https://drive.google.com/file/d/1Z0Gq04a9pl_1dJCkwlLLzeMNq_8ToUOq/view (accessed 4 February 2022).

Cameron, D. (2001) *Working with Spoken Discourse*. Sage, Thousand Oaks, CA, USA.

Capistrano, R.C. and Notorio, P.A. (2020) A content analysis of the future of tourism through the presidential state of the nation address in the Philippines (1987-2019). *Journal of Tourism Futures* 7(1), 131–146.

Castells, M. (2004) *The Power of Identity: The Information Age – Economy, Society, and Culture*, 2nd edn. Wiley-Blackwell, Malden, MA, USA.

Cheong, S.M. and Miller, M.L. (2000) Power and tourism: a Foucauldian observation. *Annals of Tourism Research* 27(2), 371–390.

Come, W.D., Schlecht, E. and Brinkmann, K. (2016) Characterization of ruminant husbandry systems in protected areas of Batanes Islands, Philippines. *Annals of Tropical Research* 38(2), 17–29.

De Leon, R.C. and Kim, S.M. (2017) Stakeholder perceptions and governance challenges in urban protected area management: the case of the Las Piñas–Parañaque critical habitat and ecotourism area, Philippines. *Land Use Policy* 63, 470–480.

Dela Santa, E. (2015) The evolution of Philippine tourism policy implementation from 1973 to 2009. *Tourism Planning & Development* 12(2), 155–175.

Dela Santa, E. and Saporsantos, J. (2016) Philippine Tourism Act of 2009: tourism policy formulation analysis from multiple streams. *Journal of Policy Research in Tourism, Leisure and Events* 8(1), 53–70.

Dela Santa, E. and Tiatco, S.A. (2019) Tourism, heritage and cultural performance: developing a modality of heritage tourism. *Tourism Management Perspectives* 31, 301–309.

Diaz-Bone, R., Bührmann, A., Rodríguez, E., Schneider, W. and Kendall, G. (2007) The field of Foucaultian discourse analysis: structures, developments and perspectives. *Forum Qualitative Social Research* 8(2), 1–27.

Farrell, T.A. and Marion, J.L. (2002) The protected area visitor impact management (PAVIM) framework: a simplified process for making management decisions. *Journal of Sustainable Tourism* 10(1), 31–51.

Fefer, J., De Urioste-Stone, S.M., Daigle, J. and Silka, L. (2018) Understanding the perceived effectiveness of applying the visitor experience and resource protection (VERP) framework for recreation planning: a multi-case study in US National Parks. *The Qualitative Report* 23(7), 1561–1582.

Foucault, M. (1974) *Prisons et Asiles Dans le Mecanisme du Pouvoir*, Vol. t.II. Gallimard, Paris.

Foucault, M. (1980) *Power/Knowledge: Selected interviews and other writings 1972-1977*. Pantheon, New York, NY, USA.

Fyfe, G. and Law, J. (1988) *Picturing Power: Visual Depiction and Social Relations*. Routledge, London, UK.

Gúčik, M. and Marciš, M. (2020) *Sustainable Tourism Indicators in Protected Areas*. Available at: https://pb.edu.pl/oficyna-wydawnicza/wp-content/uploads/sites/4/2020/06/Obszary-przyrodniczo-cenne-w-rozwoju-turystyki1.pdf (accessed 4 February 2022).

Henderson, J.C. (2011) Tourism development and politics in the Philippines. *TOURISMOS: An International Multidisciplinary Journal of Tourism* 6(2), 159–173.

Herman, A. (2016) *Visual intelligence: Sharpen your perception, change your life*. Houghton Mifflin Harcourt, Boston, MA, USA.

Hollinshead, K. (1999) Tourism as public culture: Horne's ideological commentary on the legerdemain of tourism. *International Journal of Tourism Research* 1(4), 267–292.

Ife, J. (2013) *Community Development in an Uncertain World*. Cambridge University, Cambridge, UK.

Jocano, F.L. (1997) *Filipino Value System: A Cultural Definition*. PUNLAD Research House, Quezon City, Philippines.

Joyner, L., Lackey, N.Q. and Bricker, K.S. (2019) Community engagement: an appreciative inquiry case study with Theodore Roosevelt National Park Gateway Communities. *Sustainability* 11(24), 7147.

Kristjánsdóttir, K.R., Ólafsdóttir, R. and Ragnarsdóttir, K.V. (2018) Reviewing integrated sustainability indicators for tourism. *Journal of Sustainable Tourism* 26(4), 583–599.

Maguigad, V.M. (2013) Tourism planning in archipelagic Philippines: a case review. *Tourism Management Perspectives* 7, 25–33.

Niavis, S., Papatheochari, T., Psycharis, Y., Rodriguez, J., Font, X. *et al.* (2019) Conceptualising tourism sustainability and operationalising its assessment: evidence from a Mediterranean community of projects. *Sustainability* 11(15), 4042.

Nyaupane, G.P., Poudel, S. and York, A. (2020) Governance of protected areas: an institutional analysis of conservation, community livelihood, and tourism outcomes. *Journal of Sustainable Tourism*, 1–20.

Ong, P.S. (2002) Current status and prospects of protected areas in the light of the Philippine biodiversity conservation priorities. *Proceedings of the Fourth Conference on the Protected Areas of East Asia: Benefits Beyond Boundaries in East Asia. Taiwan Organizing Committee for the Fourth Conference of the Protected Areas of East Asia*, Taipei, Taiwan.

Philippines Statistics Authority (2018). Available at: https://psa.gov.ph/content/batanes-quickstat-may-2018 (accessed 4 February 2022).

Provincial Government of Batanes (2018) *Batanes Ecotourism Master Plan*. Batanes, Philippines.

Provincial Government of Batanes (2019) *Visitor arrival, Province of Batanes, 2004-2019 (Jan-Sept)*. Batanes, Philippines.

Ramos-Tumanan, M.A. (2018) Place Attachment of Tourists and Place Meaning by Residents: Cases from Huizhou, Anhui China. Doctoral thesis. University of Waikato Management School, New Zealand.

Ramos-Tumanan, M.A. and Ryan, C. (2019) Chinese rural tourism: seeking a Chinese perspective from an emic stance, or a case of post hoc reflection? *Tourism Management* 75, 284–292.

Ryan, C. and Aicken, M. (2005) *Indigenous Tourism: The Commodification and Management of Culture*. Elsevier, Oxford, UK.

Ryan, C. and Huang, S. (2013) *Tourism in China: Destinations, Planning and Experiences*. Channel View Publications, Bristol, UK.

Ryan, C., Zhang, C. and Deng, Z. (2011) The impacts of tourism at a UNESCO heritage site in China – a need for a meta-narrative? The case of the Kaiping Diaolou. *Journal of Sustainable Tourism* 19(6), 747–765.

Ryan, C., Chen, Z., Linrui, L. and Ramos-Tumanan, M.A. (2019) Tourism planning: the United Nations World Tourism Organization INSTO Programme–the example of Raglan, New Zealand. In: Gursoy, D. and Nunkoo, R. (eds) *The Routledge Handbook of Tourism Impacts, Theoretical and Applied Perspectives*. Routledge, London, UK, pp. 62–71.

Stoker, P., Rumore, D., Romaniello, L. and Levine, Z. (2021) Planning and development challenges in Western gateway communities. *Journal of the American Planning Association* 87(1), 21–33.

Strickland-Munro, J.K., Allison, H.E. and Moore, S.A. (2010) Using resilience concepts to investigate the impacts of protected area tourism on communities. *Annals of Tourism Research* 37(2), 499–519.

Telfer, D.J. and Sharpley, R. (2015) *Tourism and Development in the Developing World*. Routledge, London, UK.

United Nations World Tourism Organization – International Network of Sustainable Tourism (2019) *Observatories: A Network of Tourism Observatories Monitoring Sustainable Tourism Development at Destination Level*. Available at: http://insto.unwto.org/ (accessed 4 February 2022).

Vandaele, J. (2016) What is an author, indeed: Michel Foucault in translation. *Perspectives* 24(1), 76–92.

Wang, C. and Xu, H. (2014) The role of local government and the private sector in China's tourism industry. *Tourism Management* 45, 95–105.

Weeks, R., Russ, G.R., Alcala, A.C. and White, A.T. (2010) Effectiveness of marine protected areas in the Philippines for biodiversity conservation. *Conservation Biology* 24(2), 531–540. DOI: 10.1111/j.1523-1739.2009.01340.x.

9 The Praia do Sono (Paraty, Brazil) as a Gateway Community to Achieve the Universalization of Sustainable Development Goal 6 (Water Sanitation) in Tourist Areas

Milena Manhães Rodrigues[1*], Renato de Oliveira dos Santos[1], Sidnei Raimundo[1] and Gustavo C.X.M.P. Machado[2]

[1]University of São Paulo, Brazil; [2]Federal University of Rio de Janeiro, Brazil

Abstract

As 4.5 billion of the world's population do not have safe sanitation, access to water and sanitation is one of the Sustainable Development Goals (SDG 6) in the 2030 Agenda. This research aims to investigate the relationship between sanitary conditions and tourism, based on ecological sanitation action research developed with the local community of Praia do Sono (Paraty, Brazil), a community gateway that integrates two protected areas: the State Ecological Reserve of Juatinga and Cairuçu Environmental Protection Area. Using a qualitative approach, specifically participant observation, during one ecological sanitation project, and semi-structured interviews with nine stakeholders, the results revealed that territorial conflicts, protected area (PA) preservationist policies and the consolidation of mass tourism has limited cultural subsistence practices and the sense of collective in the community. The water sanitation project has produced more belonging and identity in the community through the recognition of the health benefits and improved quality of life for residents.

Keywords: Sustainable Development Goals (SDGs), ecological sanitation, tourism, local community, protected areas (PAs)

9.1 Introduction

According to the World Health Organization (WHO), 2.1 billion of the world's population do not have access to safe drinking water at home, and 4.5 billion do not have safe sanitation (WHO and United Nations Children's Fund, 2017). There is a deficit in access to these rights in Brazil, especially in rural areas, due to population dispersion and distance from public urban supply and treatment networks. This further aggravates inequities between urban and rural areas, thus accentuating problems of environmental injustice. This reality damages environmental sanitation, health and quality of life.

Besides being a human right, access to WASH (water, sanitation and hygiene) is essential to public health, people's productive capacity and quality of life, and, therefore, to sustainable development. The Sustainable Development

*Corresponding author: milenamanhaes@usp.br

 Tourism Transformations in Protected Area Gateway Communities (eds S.L. Slocum *et al.*)
DOI: 10.1079/9781789249033.0009

Goals (SDGs) are divided into 17[1] goals. The access to WASH is a specific theme of SDG 6, but also contributes to achieving other sustainable development objectives as a cross-transverse topic for each of the other goals.

Activities that depend on adequate environmental conditions, such as tourism, do not fully develop in locations with a lack of treated water and sanitation. Moreover, the traditional planning of tourism in Brazil (until the 1980s) has never incorporated sanitation costs into development projects, resulting in the expense being considered an economic externality. Thus, water contamination has affected the development of tourism in places where natural heritage is a part of the main attraction, especially in areas near natural water sources (Instituto Trata Brasil, 2018). In this context, this research is aimed at investigating the relationship between sanitary conditions and tourism, based on an ecological sanitation action research project developed with the local community of Praia do Sono, a gateway community to two protected areas (PAs) and a tourist destination that integrates a large mosaic of PAs in Paraty (Rio de Janeiro, Brazil).

9.2 SDG 6 and Tourism

Agenda 2030 (United Nations, 2015) is a global action pact aimed at sustainable development, poverty eradication, equity, peace and prosperity for people and the planet. Among the goals, SDG 6 addresses the issue of water and sanitation and aims to 'ensure the availability and sustainable management of water and sanitation for all' (United Nations, 2015, p. 20). SDG 6 establishes eight targets[2] regarding the univerzalisation of access to drinking water, as well as adequate sanitation and hygiene, by linking its efficient and sustainable use to cooperation for training and management of water resources, conditional on the participation of local communities (United Nations, 2015).

To some extent, SDG 6 relates to all the other SDGs (United Nations, 2015; National Water Agency, 2019). This crossover is exemplified in *Water and Sanitation Interlinkages Across the 2030 Agenda for Sustainable Development* (United Nations, 2016), when it points to increased access to water supply, sanitation and hygiene in homes, health facilities, schools and workplaces, complemented by wastewater treatment, as a way to reduce the risk of waterborne diseases (SDG 3) and malnutrition (SDG 2); support for education (SDG 4); a productive workforce (SDG 8); reduction in poverty (SDG 1); gender inequality (SDG 5); and other inequalities (SDG 10). This chapter highlights some SDGs that, if not fully implemented as projected, may negatively impact water resources and water-related ecosystems, compromising, for example, increased agricultural productivity (SDG 2), energy security for all (SDG 7), and gross domestic product-based economic growth in the least developed countries (SDG 8) (United Nations World Tourism Organization and Pacto Mundial Red Española, 2016).

In addition to the effects of sanitation on human development, its impact on tourism is highlighted. In nature-based destinations, where environmental conditions are essential assets for tourism, water and sewage management is a crucial concern for tourism sustainability (Wells *et al.*, 2016), as tourism adds additional pressures on water and waste infrastructure. Nobel *et al.* (2012) argue that the respect for human rights, which includes access to adequate water and sanitation, must be a premise of sustainable tourism. According to the authors, if the touristic activity compromises access to these rights or puts the health and wellbeing of the community at risk, it can affect the socioeconomic mobility, livelihood, food security and the sustainability of tourism itself.

There is a direct relationship between better performing sanitation destinations and the higher international flows of tourists and job generation through tourism (Instituto Trata Brasil, 2018). Moreover, the sanitary structure of a locality serves both the tourist and the resident, where a place that supports quality of life for residents is also a good place for the quality of the tourist experience (Jafari, 2012). Therefore, it is important that tourism planning includes environmental sanitation, represented by SDG 6, as an element of sustainable development; however, this is often a peripheral issue of tourism planning (Irving, 2018).

The contributions of tourism for PAs and local development can be powerful as an avenue through which the SDGs can be accomplished.

However, according to Boluk *et al.* (2019), studies relating tourism and SDGs are incipient and often associated with those in which the activity is explicitly mentioned, for example: SDG 8 concerning the promotion of 'sustained, inclusive, and sustainable economic growth, full and productive employment and decent work for all' (United Nations World Tourism Organization (UNWTO) and Pacto Mundial Red Española, 2016, p. 22); SDG 12 aimed at 'ensuring sustainable patterns of production and consumption' (p. 23); and SDG 14 concerning 'conservation and sustainable use of oceans, seas, and marine resources for development' (p. 23). Yet, 'tourism can play a crucial role in achieving access to water and safety, as well as hygiene and sanitation for all' (p. 4), as tourism can boost investment in infrastructure, including sanitation (Frone and Frone, 2011).

As tourism increases the demand for water and sanitation services while intensifying the volume of wastewater to be treated, it can compromise the ecological systems of tourist destinations, especially in isolated areas where water treatment availability is already scarce (Lavanchy, 2017). In this context, Cole (2012) and Cole and Browne (2015), in analysing and revisiting the relationship between water (in) equity and tourism in Bali, state that the diversity of these transient users, when associated with weak governance systems, accentuates inequalities in tourist destinations and marginalizes traditional knowledge and its relationship with water. This perspective can also be considered in relation to sanitation.

Nobel *et al.* (2012) argue that challenges, such as deficiency of structure and participative management, unregulated water privatization and low levels of conscience, are, probably, faced by most touristic destinations around the world. Thus, the availability and management of environmental sanitation are challenges to global sustainable development, including sustainable tourism (Irving, 2018), and their lack is detrimental to adequate development of community-based tourism, at the local scale.

In emerging countries and peripheral regions, specifically the Global South, tourism, as a sociocultural and economic phenomenon, has been configured as an object of policy actions, investments and research, even if still in an incipient way. However, the method by which tourism has been established, historically, potentially excludes the receiving communities that, many times, have not achieved the promises disseminated in official political–economic discourses regarding job generation, income distribution, and local and regional development (Mendonça *et al.*, 2020). Moreover, this model of conventional tourism compromises cultural, environmental and structural aspects of a community, including water and sanitary issues, separating tourism development from sustainable development for the communities and their territories.

In this context, the commitment to global sustainable development demands the co-formulation, implementation and monitoring of the SDGs at the local level, understanding the territorialization of this process, and the role of social participation in bottom-up approaches, including fostering local development policies (Global Taskforce of Local and Regional Governments, United Nations Development Programme, and United Nations Habitat, 2016). This means that access to water and adequate sanitation, as well as other SDGs, involves the participatory management and alignment of actions (United Nations, 2015). In a correlated way, at the national level, the Rural Brazil Sanitation Program guides local and territorialized actions towards the universalization of sanitation as a human right, associating structural actions (works) to structuring actions (education and social participation) (Brasil Ministério da Saúde, Fundação Nacional de Saúde, 2019). Therefore, this case study approaches the Praia do Sono (Paraty, Brazil) as a gateway community to achieve the universalization of SDG 6 (water sanitation) in tourist areas.

9.3 The Community of Praia do Sono

Praia do Sono is located in one of the richest regions of biodiversity in the Atlantic Forest (Fig. 9.1), the Bocaina Mosaic. With a total of 216,000 hectares of forest, the Bocaina Mosaic integrates 18 PAs and their buffer zones, incorporating territories between the south coast of the state of Rio de Janeiro (Angra dos Reis and Paraty) and part of the north coast of the state of São Paulo (Ubatuba) (Gallo *et al.*, 2016). The Bocaina Mosaic includes the territories

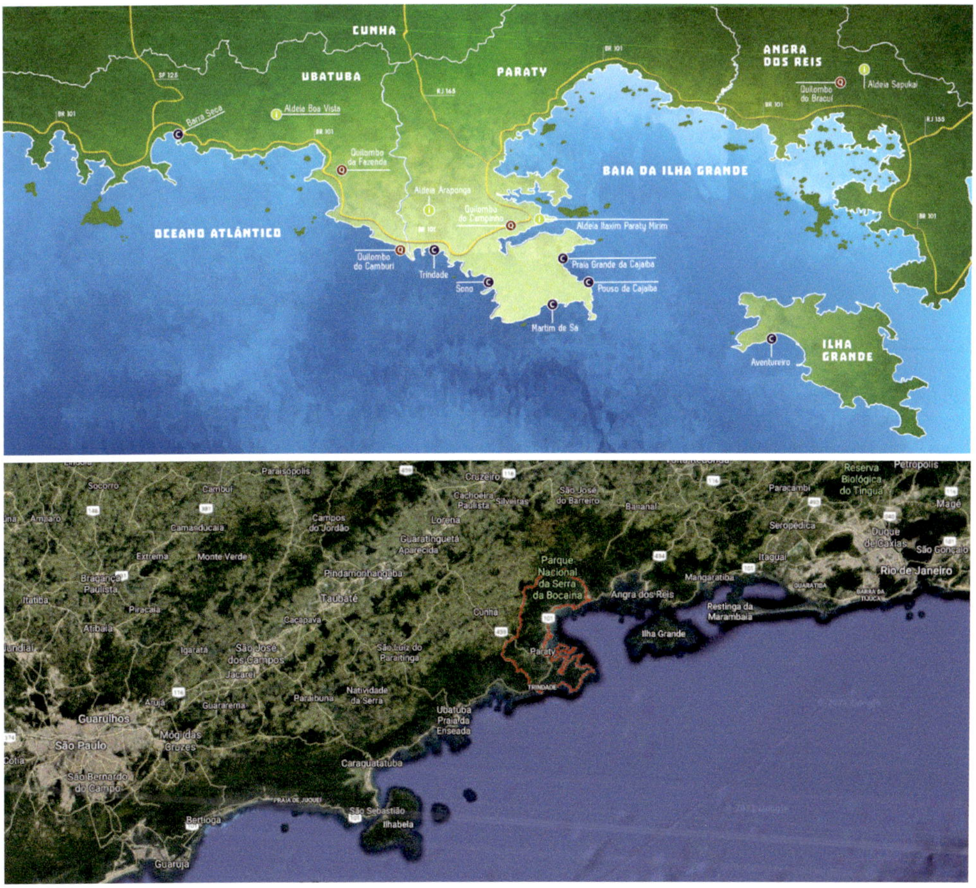

Fig. 9.1. Location of Praia do Sono (Paraty, Rio de Janeiro, Brazil) (Observatory of Healthy and Sustainable Territories of Bocaina, https://issuu.com/forumdecomunidadestradacionais/docs/folheto_tbc_fct_temporario_2015_liv).

of 51 traditional communities – such as the Indigenous Quilombolas and Caiçaras, which are historically inserted in a context of socioenvironmental conflicts, mainly related to tourism and the creation of PAs (Mendonça *et al.*, 2020). In this context, Cioffi and Raimundo (2020, p. 10) point out that there is a crisis that threatens the success of environmental policies: these communities, who live in heritage sites, demand access to, use of and participation in the management of these areas. This creates tensions between the interests of heritage conservation by the authorities and the lifestyle needs of the inhabitants of these areas.

Thus, as an alternative to resistance, these traditional communities have created, and are organized by, the Forum of Traditional Communities of Angra dos Reis, Paraty, and Ubatuba (FCT). Created in 2006, the FCT aims to defend the rights of traditional communities of this territory by: participation in public policies; affirmative agendas and supporting technological solutions that contribute to improving the quality of life; maintaining the communities in their territories; supporting natural resource management; and valuing traditional ways of life for Indigenous communities (Gallo *et al.*, 2016).

The area where the Bocaina Mosaic is located was officially declared a World Heritage Site in 2019. It is a mixed World Heritage Site (the first in Brazil), showcasing the wealth of local biodiversity associated with the way of life and cultural references of traditional communities (United Nations Educational, Scientific and Cultural Organization, 2019). Among the territories and communities that give life to this heritage is the Praia do Sono, a tourist destination that integrates two overlapping PAs: Juatinga State Ecological Reserve – REEJ (IUCN Category I) and Cairuçu Environmental Protection Area (IUCN Category VI). There are 314 Caiçaras who live in Praia do Sono, which encompasses 22% of the entire population of the REEJ (Machado, 2019). The term *Caiçara* is attributed to the traditional communities that live along the coast of the states of Paraná, São Paulo and Rio de Janeiro whose economy and culture are based on fishing and subsistence agriculture (Mendonça *et al.*, 2017)

Large infrastructure undertakings related to mass tourism, starting in the 1970s, boosted unplanned urbanization in this territory, intensified the vulnerability and uneven development in the Bocaina Mosaic, and increased the pressure on natural resources and traditional communities (Gallo *et al.*, 2016). Real estate speculation, the development of private condominiums and the restrictions established by the PAs (such as subsistence and extractive agriculture), have limited gateway resident access to traditional resources within the PAs. Moreover, the Caiçaras from Praia do Sono have modified traditional habits in order to work within the tourism industry. These jobs include restaurants and bars, kiosks, camping, inns and vacation home rentals, as well as boat transport and artisanal fishing (Gallo and do Nascimento, 2019). The result has been a reduction in traditional subsistence practices, such as subsistence farming (Machado, 2019).

The socioeconomic and cultural conditions are challenging the permanence of this community, and the sanitary conditions are an important aspect of this resistance. Until 2016, there was no water supply network in Praia do Sono, as individuals collected water directly from the Cachoeira River. Subsequently, the community improvised a 'system' of distribution by adapting hoses, until the municipal water supply and distribution network was completed. However, there is still no collective treatment system. During the high tourist season, the scarcity of water in the community is aggravated, as visitation on New Year's Eve can reach 15,000 people (El País Brazil, 2017). Regarding sewage treatment, 80.5% of the houses have cesspits, and 19.5% still discharge sewage directly into the Barra River (Gallo *et al.*, 2016).

Cole and Browne (2015) highlight the relevance of human–environmental relations, especially in sites where the use, distribution and significance of the water is fundamental to the worldview and way of living of the community. That said, sanitary interventions must also include a cohesive alternative ethnoecological approach with interactions among people and the environment; in the case of the Caiçara community of Praia do Sono, associating sanitation engineering and permaculture as interactions between theory, science and tradition that conjugated the perception of the locals and the technicians (Gallo and do Nascimento, 2019).

Faced with local health conditions, the Observatory of Sustainable and Healthy Territories of Bocaina, an action research branch of the Oswaldo Cruz Foundation (Fiocruz), the Forum of Traditional Communities of Angra dos Reis, Paraty and Ubatuba (FCT) and the National Health Foundation (Funasa), developed an ecological sanitation project together with the Caiçara community of Praia do Sono (Gallo *et al.*, 2016). Based on the principles of ecological sanitation, a closed-cycle system for water and nutrients was collectively chosen, known as an evaporation tank. This process uses the nutrients and water in the sewage for agricultural cultivation, then returns the water to the environment via evapotranspiration (Machado *et al.*, 2018). The ecological sanitation modules use plants, which absorb nutrients, as part of the ecological sanitation system. In the case of Praia do Sono, the community opted to plant banana trees. Thus, banana trees are a visible part of the ecological sanitation in the landscape, unlike other traditional systems that are underground.

Action research is a participative research method which involves problem diagnosis, solution proposition and collective action execution, in which the researcher is a facilitator and people in the investigated situation are participants with a voice and opportunity to act. It is fundamental to develop a space for dialogue, where involved actors participate in the resolution of problems with differentiated knowledge, proposing solutions and learning through practice. For this reason, the planning of the action research must be flexible and consider adaptations as functions of the circumstances and internal dynamic of the group. It is not just about resolving a problem quickly but developing the conscience of the collective in the cultural and political solutions to the problems faced by those involved (Thiollent, 2011; Machado, 2019).

The social participation of the Praia do Sono community was essential throughout the ecological sanitation project. In August 2013, technical visits were made in the community to construct a participative diagnosis that involved teaching planners about the environmental and social interrelations and the way of life of Caiçaras in their territory. That same year, a 'spoken map' was constructed, a social cartography about sanitation from the community's perspective. During the project period, between July 2014 and December 2018, several reunions with the community and social actors were made, including a sanitary inquiry, to learn about the evolving conditions of the community. The implementation process focused on the autonomy of the community, including: training and involvement of all local actors; semi-annual workshops of participative planning with the community; monthly follow-up meetings between academic researchers and locals with the diverse stakeholders involved throughout the process; monthly meetings of the construction team and the community to support collective decision-making through consensus; information-gathering and constant assessment; seminars; and educational booklets[3] for the local actors (Gallo and do Nascimento, 2019).

The project involved the construction of 11 modules of ecological sanitation, including an ecological sanitation pilot project in the school of Praia do Sono.[4] After discussions and evaluations of the pilot project, nine additional sanitation projects were implemented in local houses, as well as the headquarters of the residents' association[5] of Praia do Sono. The selection of those 11 sites was determined by the community.

These projects included education and community participation to aid in the choice of social technology and the implementation of sanitary infrastructure, including the training and hiring of locals (Gallo *et al.*, 2016), who became construction workers, as well as social mobilizers. The community workers opted for ecological sanitation, determined the location and the implementation order of each sanitation module, and kept abreast of construction progress with visits and assessment reunions with the involved institutions, including educational activities that offered training on the maintenance of the sanitation system. The community participation in the selection of this sanitation project at Praia do Sono, using sustainable technology as well as maintenance and implementation actions, is consistent with the participation recommended by SDG 6, and the guidelines of the Rural Brazil Sanitation Program (Brasil Ministério da Saúde, Fundação Nacional de Saúde, 2019).

These improved conditions and interpretive information about the improvements to the quality of life of the community members have become a unique tourist attraction within the community-based tourism of Praia do Sono (Jaeger, 2018). Thus, the relevance of this work is due, in part, to the relationship between tourism and SDG 6. The social participation in this case study culminated in the 'creation' of an interesting touristic destination. It is hoped that the results presented here can be used by gateway communities in other regions or countries that have characteristics similar to those found on Praia do Sono.

9.4 Methodological Procedure

This research employed a qualitative approach, with a case study of action research in ecological sanitation within Praia do Sono. As social participation is essential to the universalization of sanitation (United Nations, 2015), and a practice associated with the interventions in this case study, the relationship between tourism and sanitation should be understood from the stakeholders' perspective. The techniques used were: data collection by participant observation (Spradley, 1980) during the ecological sanitation project, which made it possible to identify and select the interviewees; and semistructured interviews with nine stakeholders. Interviewees were chosen based on their role in community leadership, selecting local actors or organizations

Table 9.1. Profile of the interviewees.

Participant 1	Man, community leadership and member of OTSS
Participant 2	Woman, community leadership and member of OTSS
Participant 3	Woman, community leadership
Participant 4	Man, local and constructor of sanitation module
Participant 5	Man, community leadership
Participant 6	Man, public manager
Participant 7	Woman, public manager
Participant 8	Man, public manager
Participant 9	Woman, community leader

including representatives of the municipal government of Paraty, the Observatory of Sustainable and Healthy Territories of Bocaina (in Portuguese, OTSS), and other environmental agencies. The institutions and age of each representative are not identified in order to maintain confidentiality of the interviewees, as it would be easy to identify them in a small community such as Praia do Sono. Thus, we identify the representatives of the city hall and environmental agencies as 'public managers' (Table 9.1). The community leaders were active participants in the project, as it is part of the local culture that leadership passes information to other locals as a way of public communication (Gallo and do Nascimento, 2019).

The interviews were held between June and July 2018, at the end of the sanitation project. Machado (2019) elaborated an interview script based on five classifications: Caiçara culture; relationship with water; relationship with sewage and waste; social participation; and vision of the future. Because Praia do Sono is a tourist destination, the interaction with tourism is integrated throughout all question classifications. For the production of this chapter, extracts from the interviews by Machado (2019), with a focus on the community's relationship with sanitation and tourism in the territory, were analysed. Thus, content analysis was conducted based on the key words 'tourism', 'tourist', 'tourist attraction', 'sanitation' and 'sewage'.

9.5 Results and Discussion

Based on the interpretative analysis of the interviews by Machado (2019), the narratives were reclassified into subthemes, presented below: Caiçara culture; impacts of tourism on sanitation; and impacts of ecological sanitation on tourism.

9.5.1 Caiçara culture

This category presents narratives of the Caiçara way of life, with practices related to natural and cultural changes, contextualized by new social dynamics mainly associated with tourism. As Cole (2012) and Cole and Browne (2015) point out, traditional knowledge and the non-dichotomous relationship between human and nature are linked to social (in)equities, which denotes the relevance of this category.

> We still have a great relationship with culture, with the soul of this place, with nature... because we are part of it, and so it is, today even more so, with predatory tourism, we still manage to maintain a little bit of our traditions, our regions, our beliefs, our traditional knowledge, which gradually passes from generation to generation. [...] Today, the activities are really tourism, right?
>
> (Participant 5)

The statements of the interviewees reaffirm the community bond with nature as part of the Caiçara identity and quality of life, as highlighted by Mendonça *et al.* (2017). Besides the campsites, kiosks and chalets for rent, the community of Praia do Sono lives by traditional activities, such as agriculture and fishing.

> You're inside the forest, you're eating much healthier food, because here we don't have access to a bakery, a market and none of that [...]. But it's the lack of services that, in my view, brings the quality of life here, you know?
>
> (Participant 9)

However, tourism has been changing sociocultural practices, which affects the community–nature relationship and the sense of collectiveness. The community used to construct buildings – houses included – collectively (this practice is known in Brazil as *multirão*) and used to live from self-production and trade exchanges within the informal economy. With the advent of mass tourism, traditional practices were replaced and the Caiçara culture altered.

> To make money, you either fished, sold your fish, or the flour. [...]. You used to exchange a lot there. [...]. Then people wouldn't take anything they didn't need. [...]. One thing supplied the other.
>
> (Participant 3)

> There is no longer that good thing about you doing it together, you know, worrying about the family next door. [...]. That [tourism] leaves the communities very uncharacterized.
>
> (Participant 5)

According to Participant 2, while tourism split up the community and has been responsible for cultural changes, it is also an alternative income provider, considering the restricted activities in the PA that affect traditional livelihoods. Today, tourism is the main income source, which has transformed the way of life of Caiçaras.

> The protected area that prohibited them from planting. So, they had to stop practising the culture and started dealing more directly with mass tourism. [...]. It brings people who end up bringing another culture into the community, and then ends up influencing people.
>
> (Participant 2)

> People today feel marginalized in fishing. That's why the vast majority today has moved to tourism.
>
> (Participant 6)

Therefore, territorial conflicts, allied to preservationist policies from the creation of PAs, such as the prohibition of traditional agriculture and the consolidation of mass tourism, have limited cultural subsistence practices (Gallo *et al.*, 2016; Cioffi and Raimundo, 2020; Mendonça *et al.*, 2020). In turn, this has reduced the sense of collectivism, demarcated by the symbolic distance from nature (Machado, 2019). Faced with this scenario, and including conflicts related to land tenure and real estate speculation, the community seeks strategies of cultural and territorial resistance, such as limiting the sale of houses to non-native people. These activities are supported by initiatives from the residents' association that understands, according to Participant 3, that lands are territories inherited by the community.

> So, all this impact that came through the squatter, the conservation unit, the Laranjeiras Condominium. [...]. There were several factors that generated conflicts, and that made the community stronger and to be able to stay in the place that they stayed today.
>
> (Participant 2)

The permanence in the territory requires improvements in the quality of life, which is also related to adequate sanitation. Thus, the demand by the community for sanitation generated the ecological project as an action in favour of permanence in the territory (Gallo *et al.*, 2016). This action had its origin in the community through observations in changes in the quality of life due to declining sanitary conditions, aggravated by tourism (Jaeger, 2018).

9.5.2 Impacts of tourism on sanitation

This category includes the implications of tourist activity on the sanitary conditions of the community, including the impacts generated by the increase in the number of users (tourists) on the demand for investment in adequate treatment (Cole, 2012; Cole and Browne, 2015). An indicator of this is the residents' perception of changes in the Barra River, the community's main river, considering that the water from the river is collected for domestic use and is currently polluted (Gallo *et al.*, 2016). Thus, collectively, the method for choosing houses that would benefit from the sanitation project was to prioritize the construction of sanitation modules for the upstream houses, followed by the downstream houses.

> But this system of sewage treatment, [...] I am sure that in some years the water of Barra will be rescued and everybody will be allowed to bath, the children, you got it? The tourists, because they are a salt on the game.
>
> (Participant 9)

There is an understanding that the destination of sewage – sinks – was not adequate. As Lavanchy (2017) argues, isolated communities with few residents face the problem of underdeveloped sanitary conditions, which was highlighted and enhanced by tourism activity in this study. According to Participant 2, with tourism increasing, there is an overcharge on the sanitary system.

> Everybody has always used a cesspit, which is just a hole, right?! [...] The only difference is that the system hasn't got worse. What has got worse is the amount of constructions, right?! And the amount of tourism that comes. So, I think this has made it worse, right?
>
> (Participant 8)

There were no bathrooms in Praia do Sono and physiological needs were satisfied outdoors. In the 1990s, the practices of tourists, who had bathrooms at home, motivated the construction of the first bathrooms in Praia do Sono (Machado, 2019). However, the bathrooms were constructed in backyards, outside the houses with cesspits, and therefore they were susceptible to overflowing on the ground, causing soil contamination (Machado, 2019).

> People came to know more, and have been installing bathroom[s] inside the houses, and so it is today. But then the problem of the cesspit. People doing it there, they thought that going under the floor was all right. [...]. It contaminates the soil, and it still leaks over it too.
>
> (Participant 4)

The sanitary conditions of Praia do Sono have been aggravated by tourists, resulting in demands for sanitation interventions. Specifically, the concern with sustainability was essential to tourists for better sanitary conditions, boosting investment in the sanitation projects (Frone and Frone, 2011; Machado, 2019). Thus, the ecological sanitation project in Praia do Sono combined structur*ing* actions (education and social participation) with structur*al* actions (works) (Brasil Ministério da Saúde, Fundação Nacional de Saúde, 2019). In the process of education and participation, both the community and local actors were able to understand the interaction between sanitation and other economic areas, such as tourism.

> It is important to perceive the direct and indirect impacts of sanitation in other areas of interaction, such as education, health, tourism, solidarity economy, and good living, among other aspects.
>
> (Participant 1)

9.5.3 Impacts of ecological sanitation on tourism

This category highlights the potential of the relationship between the universalization of SDG 6, with a focus on sewage treatment and its contributions to tourism, exemplified by the ecological sanitation in Praia do Sono. Ecological sanitation contributes to sustainable development, which reinforces the transversality of SDG 6 (United Nations, 2015, 2016; National Water Agency, 2019), associating sanitary conditions with health, quality of life and financial sustainability, which are related to tourism development and the sustainability of attractions (Wells *et al.*, 2016). In this context, Participant 5 comprehends that the adequate sanitation treatment contributes to ecological sustainability, such as the cleaning of the Barra River, and counters the leisure and tourism activities related to the natural attractions.

> What is cleaning the water of the Barra River? It's to bring a better quality of life to the community, it's to bring growth in tourism, this is also part of the quality of life, you know?! It's to increase the financial resources that through tourism we survive because it's going to be so beautiful, it's going to be so adequate.
>
> (Participant 9)

Moreover, as Participant 2 states, ecological sanitation has shown tourists that the Conservation Unit and the community take care of the territory. By this process the community has perceived that ecological sanitation contributes to their territory and expands the perspectives of community-based tourism (Jaeger, 2018).

> We are talking about tourism, what this sanitation brings of benefit to tourism. [...]. A tourism that will interact with the community. [...]. And I think that today, the need today ... of implementing quality tourism, right? A tourism that we say is community-based tourism, that tourism that goes not to degradation, but to cooperation.
>
> (Participant 1)

Although sanitation contributes to the sustainability of natural attractions and supports sanitary conditions, the ecological sanitation project was not designed for tourism purposes specifically. However, the sanitation project became a source of pride in Praia do Sono, to be shown as something planned and constructed by the community. Moreover, exceeding expectations of the tourism potential, the ecological sanitation modules have become attractions as part of the community-based Praia do Sono tourism route (Buck, 2020).

> That people call up wanting to do research, want to come visit, want to do the itinerary on top of ecological sanitation. [...]. You never would have imagined that a sewage treatment would become attractive. [...]. Normally, it is a place that generates bad smells, that generates negative impacts. [...]. It's something that lies beneath the earth, which is for no one to actually see. Not ours, we make a point of showing it. So, this is very differential, this is very impacting positively.
>
> (Participant 5)

Thus, tourism can stimulate the achievement of SDG 6. Also, the universalization of water, hygiene and sanitation for all can contribute to tourism development. From this perspective, tourism development can contribute to local development, when it favours quality of life and social participation. Ecological sanitation may strengthen the community gateway with life quality, contributing to the permanence of the territory and to the guarantee of natural assets for community survival, leisure and tourism (United Nations World Tourism Organization (UNWTO) and Pacto Mundial Red Española, 2016; United Nations, 2020). When associated with community-based tourism, infrastructure projects can foster a sustainable form of development, as they guarantee communitarian advocacy in the care of territories and cultures (Cole and Browne, 2015; United Nations, 2015; Buck, 2020).

9.6 Conclusion

Tourism can benefit from the sanitary infrastructure of destinations, but it can also provide investment to support SDG 6, from the perspective of the universalization of access to adequate sanitation (Frone and Frone, 2011; United Nations, 2015). In this context, Nobel *et al.* (2012) identified examples of good practice, pointing out principles for water equity that include tourism, and developed recommendations, including: government positioning in guaranteeing citizens' rights; improving water infrastructure and sewage treatment; participatory planning and management of water resources; sharing information about new technologies; establishing guidelines for tourism companies and training; and raising awareness among users, including tourists. In line with this approach, this research aimed to investigate the relationship between sanitary conditions and tourism, based on ecological sanitation action research developed with the local community of Praia da Sono (Gallo *et al.*, 2016; Machado, 2019).

This case study evidences the interaction between SDG 6 and tourism, via the implementation of ecological sanitation with the Caiçara community of the tourist destination Praia do Sono (Brazil), which fostered innovation, cooperation, social participation and learning. The research showed that community participation was an essential condition for the development of the sanitation project. The participation in the discussion arenas of FCT, Association of Residents of Praia do Sono, and OTSS, and the participative planning of concrete actions, allowed more engagement and reflection by the community. This reflection included pride in their territory and identity, and a sense of belonging, which was reflected in the improved sanitary conditions, quality of life and tourism development.

The investigation revealed that the relationship with tourism is complex: it is sometimes seen as a vector of pollution and cultural change (from the perspective of mass tourism), and sometimes as a strategy of resistance and permanence in the territory (through the discourse of community-based tourism) (Mendonça *et al.*, 2017; Buck, 2020). Sanitation, contrary to pollution, was another element of this relationship. In this case study, the narratives reported that, besides recognizing the benefits to the health and quality of life of the residents, the ecological sanitation modules started receiving visitors and became one of the community's tourist attractions, extrapolating expectations on the impacts of tourism for sanitation as well as sanitation for tourism. Moreover, considering that access to adequate sanitation involves the participatory management and alignment of actions (United Nations, 2015), associating structural actions (works) to structuring actions (education and social participation) (Brasil Ministério da Saúde, Fundação Nacional de Saúde, 2019), the sanitary interventions of Praia do Sono enhanced social participation and inclusion of traditional knowledge. The choice of social technology, the selection of new construction sites and the ability to apply input for infrastructure management, including educational activities and local maintenance of sanitary systems, all supported community enrichments that in turn support sustainable development within the realm of SDG 6. Thus, the Caiçara community from Praia do Sono may be an inspiration to other gateway communities in PAs that aim to play a leading role in their development.

Regarding the limitations of the study, we recognize that the results are not generalizable. However, we hope they provide the groundwork for additional studies in a variety of locations in an attempt to better understand the challenges and success of SDG 6 sanitation projects and their relationship to tourism. In terms of the possibilities for future research, in addition to sanitation, the discussion on SDG 6 could be expanded to include a focus on the relationship between tourism and water, and comparative studies could take place with other communities in the Bocaina Mosaic that have also developed ecological sanitation. It is also relevant to monitor the continuity of the project in Praia do Sono over time. The aspects of SDG 6 and tourism interaction described are important and can be considered in other destinations that present similar situations. The implementation of sanitation actions can improve the quality of life of local communities and tourists, alike.

Notes

[1] The 17 SDGs are divided into 17 goals: (1) No Poverty; (2) Zero Hunger; (3) Good Health and Wellbeing; (4) Quality Education; (5) Gender Equality; (6) Clean Water and Sanitation, the topic of this chapter; (7) Affordable and Clean Energy; (8) Decent Work and Economic Growth; (9) Industry, Innovation and Infrastructure; (10) Reducing Inequality; (11) Sustainable Cities and Communities; (12) Responsible Consumption and Production; (13) Climate Action; (14) Life Below Water; (15) Life On Land; (16) Peace, Justice, and Strong Institutions; (17) Partnerships for the Goals (United Nations, 2015).

[2] **Goal 6 targets**. 6.1 By 2030, achieve universal and equitable access to safe and affordable drinking water for all. 6.2 By 2030, achieve access to adequate and equitable sanitation and hygiene for all and end open defecation, paying special attention to the needs of women and girls and those in vulnerable situations. 6.3 By 2030, improve water quality by reducing pollution, eliminating dumping and minimizing release of hazardous chemicals and materials, halving the proportion of untreated wastewater and substantially increasing recycling and safe reuse globally. 6.4 By 2030, substantially increase water-use efficiency across all sectors and ensure sustainable withdrawals and supply of freshwater to address water scarcity and substantially reduce the number of people suffering from water scarcity. 6.5 By 2030, implement integrated water resources management at all levels, including through transboundary cooperation as appropriate. 6.6 By 2020, protect and restore water-related ecosystems, including mountains, forests, wetlands, rivers, aquifers and lakes. 6.A By 2030, expand international cooperation and capacity-building support to developing countries in water- and sanitation-related activities and programmes, including water harvesting, desalination, water efficiency, wastewater treatment, recycling and reuse technologies. 6.B Support and strengthen the participation of local communities in improving water and sanitation management (United Nations, 2015, pp. 20–21).

[3] For instance, the guide 'Guia: Caminhos e cuidados com as águas – Faça você mesmo seu sistema de Saneamento Ecológico, available at: https://issuu.com/otss/docs/v5_finalsiteotss_cartilha_saneament (accessed 4 February 2022).

[4] The implementation of this module may be watched on the video available at: https://vimeo.com/132495207 (accessed 4 February 2022).

[5] The residents' association is a collective of inhabitants in a neighbourhood or community, organized to centralize issues of the territory and propose social actions.

References

Boluk, K.A., Cavaliere, C.T. and Higgins-Desbiolles, F. (2019) A critical framework for interrogating the United Nations Sustainable Development Goals 2030 Agenda in Tourism. *Journal of Sustainable Tourism* 27(7), 847–864. DOI: 10.1080/09669582.2019.1619748.

Brasil Ministério da Saúde, Fundação Nacional de Saúde (2019) *Programa Nacional de Saneamento Rural*. Funasa. Available at: http://www.funasa.gov.br/documents/20182/38564/MNL_PNSR_2019.pdf/08d94216-fb09-468e-ac98-afb4ed0483eb (accessed 4 February 2022).

Buck, N.K. (2020) *Turismo de Base Comunitária no território da Bocaina - Brasil: tecendo parcerias e redes em busca da sustentabilidade*. Dissertação de Mestrado, Fundação Getulio Vargas, Escola de Administração de Empresas de São Paulo. Available at: http://bibliotecadigital.fgv.br/dspace/bitstream/handle/10438/29209/Trabalho%20Aplicado%20MPGC-%20Sustentabilidade.pdf?sequence=3&isAllowed=y (accessed 4 February 2022).

Cioffi, G. and Raimundo, S. (2020) Áreas naturais protegidas: Uma reflexão sobre a patrimonialização da natureza pela sociedade ocidental. *Caderno Virtual de Turismo* 20(2), 1–18. DOI: 10.18472/cvt.20n2.2020.1851.

Cole, S. (2012) A political ecology of water equity and tourism: a Case Study from Bali. *Annals of Tourism Research* 39(2), 1221–1241.

Cole, S. and Browne, M. (2015) Tourism and water inequity in Bali: a social-ecological systems analysis. *Human Ecology* 43(3), 439–450. DOI: 10.1007/s10745-015-9739-z.

El País Brazil (2017) *The Controversial Condominium in Paraty that Created Exclusive Beaches for its Billionaires*. R. Daflon (Public Agency). Available at: https://brasil.elpais.com/brasil/2017/11/29/politica/1511987264_462758.html (accessed 4 February 2022).

Frone, S.M. and Frone, D.F. (2011) Water supply and sanitation development as a requirement for sustainable tourism development in Romania. *Recent Researches in Tourism and Economic Development*, 464–469.

Gallo, E. and do Nascimento, V. (2019) *O território pulsa: territórios sustentáveis e saudáveis da Bocaina: soluções para a promoção da saúde e do desenvolvimento sustentável territorializados*. Fiocruz, Paraty, Rio de Janeiro, Brazil.

Gallo, E.S., Ruprecht, A.F.F., Sobrinho, T., Finamore, F.X., Shubo, T.P, *et al.* (2016) Territorial solutions, governance and climate change: ecological sanitation at Praia do Sono, Paraty, Rio de Janeiro, Brazil. In: Leal Filho, W., Azeiteiro, U., and Alves, F. (eds) *Climate Change and Health. Climate Change Management*. Springer, London, UK, pp. 515–532.

Global Taskforce of Local and Regional Governments, United Nations Development Programme, and United Nations Habitat (2016) *Roadmap for Localizing the SDGs: Implementation and Monitoring at Subnational Level*. ONU, New York, NY, USA. Available at: https://www.uclg.org/sites/default/files/roadmap_for_localizing_the_sdgs_0.pdf (accessed 4 February 2022).

Instituto Trata Brasil (2018) *Economic and Social Benefits of Sanitation Expansion in Brazil*. Available at: http://www.tratabrasil.org.br/images/estudos/itb/beneficios/sumario_executivo.pdf (accessed 4 February 2022).

Irving, M.A. (2018) Participação, inclusão social, e os Objetivos de Desenvolvimento Sustentável: por uma nova ética em projetos turísticos. In: Irving, M.A., Azevedo, J. and Lima, M.A.G. (eds) *Turismo: Ressignificando sustentabilidade*. Folio Digital, Portugal, pp. 57–89.

Jaeger, M.A.H. (2018) Sonhando o Sono: Ecoturismo e protagonismo comunitário na Praia do Sono (Paraty/RJ). Dissertação de Mestrado Universidade Federal do Estado do Rio de Janeiro. Available at: http://www.unirio.br/ccbs/ecoturismo/arquivos/dissertacoes/turma-2016/MayaraJaeger_TCC.pdf (accessed 4 February 2022).

Jafari, J. (2012) *Global Summit on City Tourism. Catalyzing Economic Development and Social Progress, 14 -16 November 2012*. Istanbul, Turkey. Summit Summary Report and Recommendation. Available at: https://webunwto.s3-eu-west-1.amazonaws.com/imported_images/36684/summit_summary_report_istanbul_2012_0.pdf (accessed 4 February 2022).

Lavanchy, G.T. (2017) When wells run dry: water and tourism in Nicaragua. *Annals of Tourism Research* 64(3), 37–50. DOI: 10.1016/j.annals.2017.02.006.

Machado, G.C.X.M. (2019) Saneamento Ecológico: Uma Abordagem Integral de Pesquisa-ação aplicada na Comunidade Caiçara da Praia do Sono em Paraty. Tese de Doutorado, Universidade Federal do Rio de Janeiro. Available at: http://pos.eicos.psicologia.ufrj.br/wp-content/uploads/2019_Dout_Gustavo_Carvalhaes-Xavier.pdf (accessed 4 February 2022).

Machado, G.C.X.M.P., Maciel, T., Carrera, L. and Gallo, E. (2018) Environmental educommunication and ecology of knowledge in the Caiçara Community of Praia do Sono, Paraty, RJ, Brazil. *The International Journal of Sustainability Policy and Practice* 13(4), 15–31. DOI: 10.18848/2325-1166/CGP/v13i04/15-31.

Mendonça, T.C.M., Santos, R.O., Lopes, P.C.B., Andrade, S.R. and Moraes, A.P.V. (2017) Turismo de Base Comunitária na Costa Verde (RJ): caiçaras, quilombolas e indígenas. *Revista Brasileira de Ecoturismo* 10(2), 328–356.

Mendonça, T.C.M., Santos, R.O. and Leite, M.C. (2020) Rede Nhandereko de Turismo de Base Comunitária: a Resistencia caiçara, quilombola e indígena tece um novo mapa de turismo na Costa Verde (RJ). In: dos Santos Alves, K. (ed.) *Diálogos Sociais em Turismo: Elementos hegemônicos e contra hegemônicos*. Dialética, Switzerland, pp. 163–180.

National Water Agency (Brazil) (2019) *(ANA). SDG 6 in Brazil: ANA's Vision of the Indicators*, Translated by Tathiana Abreu. ANA, Brasília. Available at: https://www.ana.gov.br/noticias/ods-6-ana-lanca-publicacao-inedita-para-acompanhar-metas-para-agua-e-saneamento-ate-2030-1/ods6_en_digital.pdf (accessed 4 February 2022).

Nobel, R., Browne, M., Cole, S., Latchford, R., Nang, D, *et al.* (2012) *Water Equity in Tourism: A Human Right - A Global Responsibility*. Tourism Concern, London, UK. Available at: https://tourismconcern.org.uk/wp-content/uploads/2014/10/Water-Equity-Tourism-Report-TC.pdf (accessed 4 February 2022).

Spradley, J.P. (1980) *Participant Observation*. Holt, Rinehart and Winston, New York, NY, USA.

Thiollent, M. (2011) *Metodologia da Pesquisa-ação*, 18. Cortez, São Paulo, Brazil.

United Nations (2015) *Transforming our World: the 2030 Agenda for Sustainable Development*, Resolution Adopted by the General Assembly on 25. United Nations, Geneva, Switzerland. Available at: https://sustainabledevelopment.un.org/content/documents/21252030%20Agenda%20for%20Sustainable%20Development%20web.pdf (accessed 4 February 2022).

United Nations (2016) *Water and Sanitation Interlinkages Across the 2030 Agenda for Sustainable Development*. Geneva, Switzerland. Available at: https://www.unwater.org/publications/water-sanitation-interlinkages-across-2030-agenda-sustainable-development/ (accessed 4 February 2022).

United Nations (2020) *Water, Sanitation and Hygiene*. Available at: https://www.unwater.org/water-facts/water-sanitation-and-hygiene/ (accessed 5 November 2020).

United Nations Educational, Scientific and Cultural Organization (2019) Paraty and Ilha Grande become the first mixed World Heritage site located in Brazil. Available at: http://www.unesco.org/new/pt/rio20/singleview/news/paraty_becomes_the_first_mixed_world_heritage_site_in_brazil/ (accessed 5 November 2020).

United Nations World Tourism Organization (UNWTO) and Pacto Mundial Red Española (2016) El Sector Turístico y los Objetivos de Desarrollo Sostenible: Turismo responsable, un compromiso de todos. Available at: https://www.e-unwto.org/doi/pdf/10.18111/9789284418299 (accessed 31 January 2022).

Wells, E.C., Zarger, R.K., Whiteford, L.M., Mihelcic, J.R., Koenig, E.S. *et al.* (2016) The impacts of tourism development on perceptions and practices of sustainable wastewater management on the Placencia Peninsula, Belize. *Journal of Cleaner Production* 111(12), 430–441. DOI: 10.1016/j.jclepro.2014.08.050.

World Health Organization and United Nations Children's Fund (2017) *Progress on Drinking Water, Sanitation and Hygiene*. Updates and SDG Baselines. Available at: https://www.who.int/mediacentre/news/releases/2017/launch-version-report-jmp-water-sanitation-hygiene.pdf?%20ua=1 (accessed 4 February 2022).

10 Public–private Partnerships: A Framework for National Park Gateway Community Development

Ole R. Sleipness, Jake Powell and Kynda R. Curtis*
Utah State University, Utah, USA

Abstract
Planning and development in national parks and their gateway communities often transcend jurisdictional boundaries. Due to their interrelated nature, addressing development needs requires increased levels of coordination between public land management agencies, private landowners and community stakeholders. This chapter describes how a public–private partnership (PPP) was leveraged to provide design solutions (structures, infrastructure and open spaces) for a proposed visitor centre and amenities at Zion National Park's eastern gateway. Proposed infrastructure and visitor experiences reflect the culture and traditional industries of the area, incorporating agricultural and food elements. The public–private partnership facilitated the needed development at Zion's eastern gateway while, simultaneously, preserving cultural and historical aspects of the space. Considerations on the effectiveness of public–private-partnerships in gateway community contexts are also discussed.

Keywords: Zion National Park, public–private partnerships (PPPs), agriculture heritage, design, gateway communities

10.1 Introduction

The US National Park Service (NPS) has been called 'America's best idea' (Weber and Sultana, 2013) and is a source of pride for most Americans. National parks exemplify many of American society's values, such as democracy and freedom. In quoting Theodore Roosevelt in 1903, Stacy Allen (2012, p. 1), former President of the Association of National Parks, stated:

> the principle behind Yellowstone and the other national parks 'is its essential democracy'. Citizens were securing 'to themselves and their children' through the landmark political actions of their elected or duly appointed representatives, the enjoyment of the parks 'in perpetuity ... by assuming ownership in the name of the nation ... for the people as a whole instead of leaving the enjoyment thereof to be confined to the very rich or select few'.

Today, there are 411 protected areas (PAs), including 84 million acres, located in every US state, the District of Columbia, American Samoa, Guam, Puerto Rico and the US Virgin Islands (NPS, 2016). These PAs include not only areas of abundant wildlife and unique ecosystems, but cultural and historical monuments, battlefields, lakeshores and seashores.

*Corresponding author: kynda.curtis@usu.edu

DOI: 10.1079/9781789249033.0010

The initial NPS development model consisted of the removal of local inhabitants, the prohibition of extractive economic activities, such as hunting, mining and forestry, and the preservation of landscapes in their 'native' condition (Weber and Sultana, 2013). These practices have often placed park officials at odds with local communities, resulting in long-standing conflicts that continue to impede development due to NPS's focus on environmental resilience, rather than sustainability which includes the 'triple bottom line' of economic, social and environmental durability (NPS, 2017). The isolationist structure of national parks, both as environmental enclaves and socially detached administrative units, has limited the effectiveness of national parks in the overall management of PAs and left national parks vulnerable to public criticism (Gimmi *et al.*, 2011).

National parks provide enormous economic opportunities for rural areas. For example, in 2015 over 307 million recreation visits resulted in US$16.9 billion in visitor spending. The total economic contribution to the US (direct, indirect and induced) supported 295,000 jobs, $11.1 billion in labour income and $32 billion in economic output. Over 31% of visitor spending went to hotels, motels and bed-and-breakfasts, while over 20% went to restaurants and bars, and almost 12% to gas and oil (Thomas and Koontz, 2016). However, tourism is prone to economic leakages through the corporate ownership of tourism resources and the importation of tourism commodities, such as food and souvenirs (Everett and Slocum, 2013); thus the true economic impact of national park visitation on local communities is difficult to ascertain. Local sourcing could reduce these leakages and keep a high portion of the economic contribution for gateway communities and local agricultural production.

The NPS provides some of the largest tourism attractions in rural areas in the US. However, the NPS's primary responsibility lies within the park boundaries, leaving local communities to struggle with rapid infrastructure changes, swelling development, increasing inflation and the loss of traditional industries and open space (Frauman and Banks, 2011). Increasingly, development is forcing open space conversion into second homes and other tourist amenities, such as resorts and golf courses. Not only does this change the culture and amenities of these rural areas (Hamin, 2001), it may also reduce heritage tourism opportunities. Finding avenues to preserve traditional culture and industries is important to the social and economic sustainability of nearby communities (Finnessey, 2012).

Gateway communities – communities that lie in close proximity to national parks and whose residents are affected by NPS policies (NPS, 2006) – face a number of economic and environmental challenges. Frauman and Banks (2011) found that rapid development had contributed to increased costs of living, a reduction in the number of farms/ranches and the loss of open space outside the park boundaries as a result of national park tourism development along the Blue Ridge Parkway in North Carolina. Hamin (2001) claims that the residents in Estes Park, Colorado (gateway to the Rocky Mountain National Park) are comprised of insiders, those who are long-time residents and dependent on tourism, and outsiders who are developers, telecommuters or natural-amenity driven migrants, in the form of second-home owners and retirees, who are not dependent on tourism receipts. The insiders are increasingly being pushed away from gateway communities, as they often lack the finances, knowledge and skills to adjust to the new tourism economy.

Additionally, the tourism and hospitality industries are known for their overuse and abuse of local resources: 75% of all environmental impacts from resorts and tourism operations are a result of excessive consumption and emission of pollutants (Bohdanowicz, 2006). Media stories and previous literature have documented the long-term negative economic, social and environmental impacts of tourism on gateway communities (Finnessey, 2012; Wilcox, 2015). While the negative impacts are numerous, the primary effects stem from the enormous amount of waste generated by resorts, which leads to costly waste management schemes, increased pests and endangerment to wildlife (Kasim, 2006; Sealey and Smith, 2014).

In 2012, the NPS opened a Sustainable Operations and Climate Change Division, whose initial task was to establish the Green Park Plan to support more sustainable park operations. As part of the Green Park Plan, the NPS issued a new mandate in 2013 that encourages concessionaires to purchase a 'significant' amount of their inputs and supplies locally. The impact of this mandate on local communities could be significant, as the NPS serves more than 23.5 million customers each year at over 250 food and beverage operations operated by concessioners in 75 parks (Vogel, 2013). The success of this programme has been mixed, as each park works with different concessionaires and faces a variety of different environments and constraints (Slocum, 2017).

The NPS has struggled with poor community relationships as it has traditionally used top-down management strategies (Slocum, 2017). This policy environment is changing in response to changing cultural and social pressures from the American public and the NPS's new policy initiative to promote gateway community engagement and partnerships. However, implementing these new community-friendly policies has been shown to be sporadic, and the NPS lacks expertise to support development in gateway communities (Slocum, 2017). Hence, NPS officials are seeking support in understanding the operational challenges of new pro-local policies at the park level and to establish better community relationships in gateway communities. Recently developed public–private partnerships (PPPs) between the NPS, state agencies, community stakeholders and landowners focus on the unique needs and challenges faced by national park gateway communities undergoing economic development to enhance economic and environmental resilience.

This chapter illustrates how PPPs can facilitate needed development in national park gateways, while preserving cultural and historical aspects of the space, and providing business and employment opportunities for local residents. Specifically, the chapter provides a case study of how a PPP was leveraged to provide design solutions (structures, infrastructure and open spaces) for visitor area development at Zion National Park's eastern gateway. Additionally, considerations for effective PPPs in gateway community contexts are discussed.

10.2 Zion National Park – A Case Study

10.2.1 Interdependence of gateway communities and national parks

Gateway communities have been variably defined within the literature as those that visitors must pass through to access large holdings of public lands (Howe *et al.*, 2012), those having significant natural amenities (Deller *et al.*, 2001) and, more recently, as 'small rural towns and cities near major public lands, lakes, scenic rivers, and other notable natural amenities' (Stoker *et al.*, 2021, p. 22). While the presence and proximity of national parks is traditionally viewed as an economic asset for adjacent communities, the value of these gateway communities to national parks is increasingly apparent (Stoker *et al.*, 2021). The inherent interdependence between communities and parks creates unique challenges for gateway communities and public land management agencies, especially when planning is uncoordinated (Howe *et al.*, 2012; Stoker *et al.*, 2021). Parks often need gateway communities to externalize growth and visitation, as parks have limited capacity to absorb all visitor impacts within their own boundaries due to natural and topographic constraints, sensitive ecological and cultural features and statutory mandates. As a case study, Zion National Park and its surrounding gateway communities are struggling to progressively address these challenges. Proactive management of visitation impacts has benefited from partnerships with private landowners adjacent to the park's eastern gateway. This coordinated approach – centred around PPPs, provides a juxtaposition of historic and contemporary efforts to develop national park gateways and the communities that support them.

10.2.2 Tensions between protection mandates and externalized visitor impacts

The 1916 National Park Service Organic Act mandates preservation and protection of parks' scenic, historic and natural features, while providing for their enjoyment – without compromising these resources for the enjoyment of future generations (Winks, 1996). Subsequent statutes, including the 1935 Preservation of Historic Sites Act, Wilderness Act of 1964, National Historic Preservation Act of 1966 and National Environmental Policy Act of 1969, constrain additional development within national park boundaries and limit adaptation to increased visitation, particularly where changes may impact natural, cultural and historic resources (Winks, 1996). Consequently, as park visitation has dramatically increased, visitor accommodations, employee housing, waste management facilities, transportation infrastructure and the evolving desires of park visitors for a broader range of recreation experiences are often externalized beyond park boundaries and borne by adjacent communities. Gateway communities typically support hotels, restaurants, guide shops, gift shops, gas stations and other services necessary for visitors, as well as the local permanent and seasonal employee workforce. As parks host a greater number of international and domestic visitors, gateway communities are no longer incidental, 'pass-through portals' to the national parks but have become an integral part of the visitor experience.

10.2.3 Uneven visitation equals uneven impacts

Nationally, visitation levels and their associated impacts are uneven across all 62 named national parks and 423 designated sites within the National Park system (NPS, 2021b). In 2020, 25% of all recreation visits occurred in the top six most-visited parks. These disparate visitation levels are impacted by park location, proximity to major metropolitan areas, convenience for international travellers and visitor interest. For example, the most visited national park, Great Smoky Mountains, is located between major population centres in the southeastern US, placing it within convenient reach of its 12.1 million visitors. In contrast to parks situated near urban areas, Zion National Park, the third most-visited, received 3.6 million visitors the same year.

Even within individual national parks (such as Zion), visitation and impacts are not evenly borne by all gateway communities surrounding and supporting the park. For example, of its 3.6 million visitors in 2020, only 1 million entered the park from its east entrance (Zion National Park Forever Project, 2021). From 2018 to 2020, the east entrance accounted for 22% of traffic volume, while the Kolob Terrace entrance accounted for only 5%. All remaining traffic entered the park through the southwestern entrance, adjacent to Springdale, Utah (IRMA Portal, 2021). Differences in visitation levels are impacted by access to primary highways, availability of community-level visitor amenities and proximity to commercial transportation, including air travel. Often, the infrastructure and development conducive to visitation were developed organically over an extended period of time, rooted in human settlements that existed prior to park establishment.

Influenced by the region's geography and major transportation infrastructure, Zion's main entrance was established at its southwestern edge, near Interstate 15, St George, and Las Vegas's international airport (Fig. 10.1), resulting in the adjacent community of Springdale becoming the park's primary gateway (Fig. 10.2), including expanded highway capacity which provides connectivity to the park. Subsequently, the park's visitor centre, employee offices and support structures were located on its more buildable southern half (Fig. 10.3). Zion's eastern gateway remained inaccessible to automobile travel through the park until a series of tunnels was constructed in 1920. The park's rugged eastern half remains less developed, except for remote trails and vehicular traffic on State Route 9, designated as the Zion National Park Scenic Byway.

The natural barriers that inhibited early development have posed challenges for contemporary regional travel. As the park initiated a shuttle system to manage burgeoning in-season visitors to main park destinations, the tunnels and steep incline of State Route 9 prevented shuttle travel from reaching the park's eastern gateway (Figs 10.4 and 10.5). While Kane County supports landowner-initiated development at the park's eastern edge, the park is both topographically and statutorily constrained and unable to significantly expand facilities within its own boundaries to service its eastern gateway. Consequently, while the park's southwestern

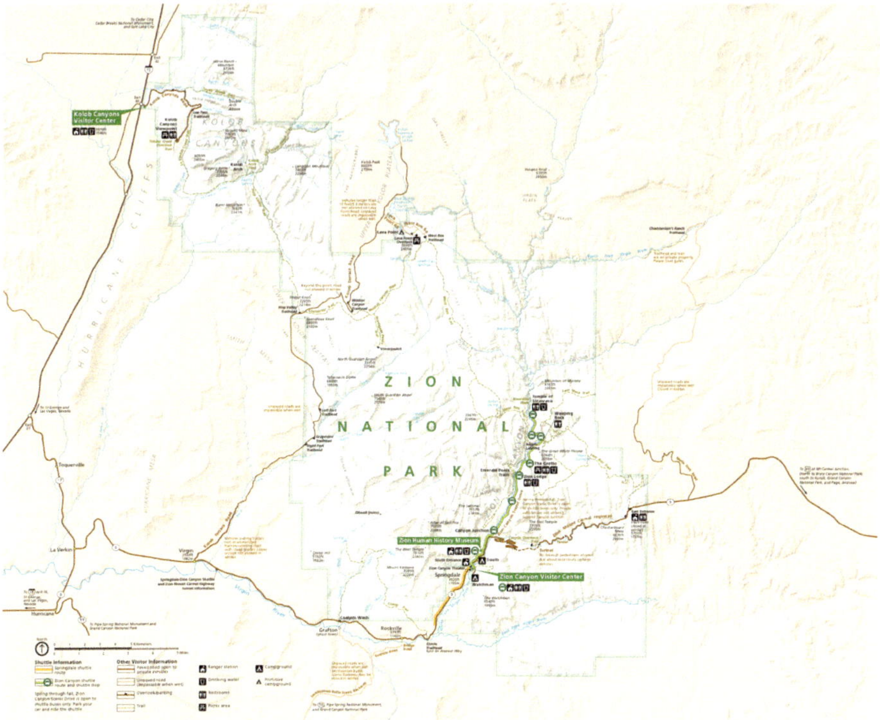

Fig. 10.1. Zion National Park's primary entrance is via the southwest gateway community of Springdale, east of Hurricane. The eastern edge of the park remains relatively undeveloped (US National Park Service, restoration/clean-up by Matt Holly, US National Park Service, 2013, http://npmaps.com/wp-content/uploads/zion-map.jpg).

edge is heavily developed, commercialized, subdivided and trafficked, its eastern gateway retains much of its early agricultural visual character, is largely unsubdivided and is owned by a small number of large landowners. Directly adjacent to the park's eastern boundary are large tracts of private land that currently support low-intensity tourist accommodations and agricultural production, farm-to-table dining and land suitable for development of additional visitor amenities (Fig. 10.6).

10.3 A New Gateway Community: A Challenge and Opportunity

Zion National Park currently has limited capacity to absorb additional visitors at its southwestern entrance. To alleviate overcrowding on its southwestern gateway, expanding visitor capacity at the eastern gateway is a possible solution. However, current park visitor amenities at the eastern entrance are limited to a single fee station and pit toilets, constructed in 1965. Reallocating visitors to its eastern edge requires developing a significant amount of infrastructure, a new visitor centre and other amenities, including shuttle connectivity to the eastern gateway. Conversations about the eastern entrance to Zion National Park being a strategic approach to address the growing visitation at Zion National Park have been occurring, both within the park and the communities that surround it, for decades. However, little progress had been made, despite the growing list of impacts due to increased visitation. In recent years, conversations have led to collaborative partnerships and a suite of proposed developments.

Land ownership surrounding the eastern entrance is limited to a small handful of

Fig. 10.2. Zion National Park's southwest gateway community, Springdale, supports a broad range of tourism development (author's own image).

Fig. 10.3. Zion National Park's existing visitor centre is located within walking distance of commercial development in its southwest gateway community of Springdale (author's own image).

Fig. 10.4. During peak season, Zion's popular main canyon is accessible only via shuttle (author's own image).

Fig. 10.5. From Zion's main canyon, State Route 9 climbs in elevation, travels through a series of tunnels, and winds through a rugged landscape to the park's less developed eastern gateway (author's own image).

Fig. 10.6. The proposed visitor contact centre site near Zion's eastern gateway is located on what is currently private land along State Route 9 (author's own image).

landowners, providing a manageable group of stakeholders when exploring expansion opportunities. Private landowners interested in the potential for expansion began working with NPS park management, Kane County, the Bureau of Land Management (BLM) and collaborated with Zion National Park's non-profit partner, Zion Forever, to explore the opportunities and scope of the potential investments at the East Zion entrance. These conversations occurred sporadically over the course of several years with gradual agreement regarding the need for facilities, such as a visitor centre, visitor accommodations, parking and transportation options rising to the top of the group's collective list of needs (Utah State University LAEP, 2018; Shoup, 2019; Zion National Park Forever Project, 2021).

While Springdale's commercial and tourism-oriented urbanization occurred organically and incrementally over the course of a century, this development has created a distinctive gateway experience for visitors accessing the park through the southwestern entrance. Additional development at Zion's eastern gateway has the potential to, in a comparatively short time frame, transform this area. This additional gateway may offer a different visitor experience framed by the area's endemic agricultural landscape. Approaching the development of a national park gateway and guiding the future development of a potential gateway community has inspired PPPs to help facilitate the process.

10.3.1 Agency coordination through a PPP approach

According to Grimsey and Lewis (2002, p. 108), PPPs are agreements in which 'public sector bodies enter into long-term contractual agreements with private sector entities for the construction or management of public sector infrastructure facilities by the private sector entity, or the provision of services using infrastructure facilities by the private sector entity to the community on behalf of a public sector entity'. Within the realm of the built environment, PPP arrangements often include construction and maintenance of energy, communications and transportation infrastructure (Kwak *et al.*, 2009), in construction projects (Tang *et al.*, 2010) and in redevelopment of vacant urban properties (Chin, 2021) where coordination between private industry and public agencies is key to timely project success. Federal statutes, such as the Federal Land Policy and Management Act of 1976, require federal land management agencies to coordinate land management plans with state and local governments (Department of the Interior, Bureau of Land Management, and Office of the Solicitor, 1976).

Federal and state government agencies have also employed PPPs to invite the private sector to maintain visitor amenities within state parks (Gilroy *et al.*, 2013), Forest Service campgrounds and national parks. Specifically, public transportation shuttles in national parks (Daigle, 2008) are provided through coordinated agreements between private concessionaires, NPS, local philanthropic partners, such as The Friends Alliance and its official non-profit partner, the National Park Foundation, which raises private funds for the benefit of NPS activities and services (NPS, n.d.). In the case of Zion National Park's eastern gateway, involvement included Kane County, Zion Mountain Ranch (ZMR), Zion National Park, Zion Forever, Five County Council of Governments, several professional consulting firms as well as non-profit consultants including Utah State University's Landscape Architecture and Environmental Planning Program.

10.3.2 Kane County's role

The Kane County, General Plan (2018) takes a position similar to other counties in rural Utah when it comes to land use and development – defining the supportive role of the county and emphasizing the rights of private property owners to make use of their land in ways that are 'within the constraints presented by the land itself, as well as traditional uses, customs, culture and available public services'. However, this apparent freedom is tempered by the County's warning that harkens to the very western mindset of the area that 'Kane County will not unduly regulate private land use, nor will public funds be used to bail out those who make unwise land use decisions' (Kane County, General Plan, 2018, p. 2).

In its General Plan, Kane County specifically states that 'unincorporated areas of the county cannot be expected to accommodate municipal-scale land uses' (2018, p. 10). However, the General Plan clarifies the county's stance that landowners are able to develop, as long as they follow the county's land use and zoning ordinances. The General Plan further explains that the county will not provide or maintain infrastructure to support private development, explicitly placing the responsibility and expense of financing, installing and maintaining these assets on the developer or landowner. Exceptions to this rule occur when development is located within a special service district. In such cases, the county is allowed to assist in the development of 'municipal-scale services' (Kane County, General Plan, 2018, p. 26). East Zion Special Service District is identified within the General Plan as one of 11 such districts located throughout the county.

In contrast to individualistic tones regarding private developers' responsibility with regards to infrastructure, Kane County's General Plan includes additional guidance on transportation corridors, explaining that the county should create safe development along all county highways. This includes the Zion Park Scenic Byway – State Route 9 from Mt Carmel Junction, through Zion National Park to Interstate 15 in cooperation with the Utah Department of Transportation and the Federal Highway Administration. The General Plan reinforces the county's perspective that land use planning and transportation should be coordinated to maximize land development opportunities. The Kane County General Plan provides neither strict restrictions nor definitive guidance with regard to development near the eastern gateway of the park. However, the plan's general approach paves the way for development, as long as it follows the broad guidelines with regards to infrastructure and respects the requirements of the larger transportation network.

To facilitate economic development, Kane County commissioned a transit study to explore scenarios for connecting Zion's southwestern and east entrances with a shuttle service, and county-wide shuttle connectivity to other established communities. The study proposed development of an electric shuttle fleet to travel State Route 9's steep eastern incline in order to access the park's eastern edge, and connections to Kane County's other communities, including Kanab.

10.3.3 Utah State University student design involvement, process, and products

In 2018, the Five County Council of Governments invited the Utah State University Landscape Architecture and Environmental Planning (USU LAEP) Program to discuss how the university might engage this contemporary design issue as a service-learning opportunity. As a 'land grant' institution (public universities established in the United States to provide primarily technical, agricultural and applied teaching and research), the USU LAEP Program employs a design engagement approach to address local and regional design dilemmas and assist public agencies, communities and non-profit organizations throughout the Intermountain West. Through 'service-learning' (Thering and Chanse, 2011), students engage a variety of diverse constituents and experience ways in which design can address tangible problems by transforming existing circumstances into preferred conditions (Simon, 1988), while developing core professional values and competencies of students as they enter the planning and design professions (Sleipness *et al.*, 2019).

In the process, USU LAEP students engaged one of the major private landowners, ZMR, in an effort to develop conceptual design alternatives for specific areas of the eastern gateway. The Program's recreation design and open space planning studio course was selected as the appropriate vehicle for providing students with this unique opportunity. Students were introduced to the project through a series of landowner meetings and a three-day immersive site visit, during which students and faculty stayed at ZMR, familiarized themselves with the landscape's unique characteristics through exploration, and evaluated potential sites for their suitability for supporting future amenity development.

During this immersive experience, students also met with local stakeholders, including other landowners, area employees and residents, representatives from Kane County government, the Zion Forever Project, and professional design firms. Students studied other national park gateway communities to better contextualize potential solutions and pitfalls when externalizing national park impacts. This collaborative process and client interaction provided a guiding framework for students to propose conceptual alternatives to the primary landowner within the stakeholder group. Throughout the process, students grappled with the challenge of creating design solutions that balanced sustainable long-term solutions with the natural conditions of the property, transportation infrastructure needs and a complicated list of required and requested design elements.

The studio design process and the students' resultant work added another layer of cooperation and a space to explore potential design opportunities as well as highlighting additional constraints. The process provided stakeholders with an additional, more exploratory, design process that was detached from the limitations often inherent in a fee for service design process. Students were free to explore and propose unconventional design solutions that advocated for the east Zion entrance to be low impact, agriculturally focused and environmentally sensitive.

Through the students' exploratory design efforts, the potential for localized food production, as a thematic design element that respected the agricultural heritage of the area, was strengthened at the site level. For example, students explored the potential for the east Zion entrance to incorporate farmers' markets, community supported agriculture, permaculture and local farm-to-fork processes and spaces not only into the landscape, but in the overarching transportation and built environment infrastructure. Moreover, because the proposed visitor centre (now known as Apple Cross Station), transit shuttle stops and shuttle parking are sited within a small valley that was historically used for agriculture (and once supported fruit production), students' design proposals included integrating food production into the visitor experience (Utah State University LAEP, 2018; Sleipness *et al.*, 2020). At the granular site level, herbs would be planted near an agrarian-styled water spigot, so hikers could pick and add mint leaves to their water bottles. Near the shuttle stop, visitor parking was broken into multiple smaller parking areas, interspersed and shaded with fruit trees, so that visitors could experience the orchards while walking to the visitors' centre or waiting for their shuttle.

While integrating agricultural activities at the site supports a distinct visitor experience, the corresponding agricultural aesthetic also supports Kane County's desired visual quality for motorists on State Route 9, which is state designated as the Zion Park Scenic Byway. In its General Plan (2018, p. 24), Kane County states its desire 'to preserve the aesthetic value of its significant scenic corridors by using the concept of visual corridor mitigation'. To minimize disturbance of its scenic corridor, the Plan encourages mitigation measures including 'using less intrusive buildings, using structures that blend in with the environment, and utilizing the natural environment to the best extent possible. The goal is to preserve the scenic value of Kane County's natural heritage – parts that may only be seen from the inside of a vehicle – but are still marketed to the world and popular to see' (2018, p. 24). In consideration of the corridor's scenic value, design proposals by USU LAEP students, as well as those prepared by professional design firms, employed an agricultural aesthetic to mitigate the visual disruptions of the proposed development, parking and transportation infrastructure.

As the Utah Department of Transportation (UDOT) has jurisdiction for ingress and egress from State Route 9 (Fig. 10.7), design proposals had to be approved by UDOT. UDOT has been less receptive to proposals for roundabouts on its state highways and, generally, reducing highway travel speeds (UDOT, 2013); however, the design students were willing to challenge UDOT's approach and proposed a roundabout where future visitor centre traffic will intersect State Route 9. Because students are able to propose less conventional solutions than professional practitioners, student design proposals sometimes facilitate discussions that might not otherwise occur, including consideration of a broader range of potential solutions (Sleipness *et al.*, 2020). At this writing, UDOT has approved proposals for the roundabout.

10.4 Project Outcomes and Developments

The conceptual design proposals that resulted from immersive on-site design activities have fuelled collaborative discussions between local landowners, agencies and stakeholders on potential design scenarios. By integrating local food production as a key design principle, proposals have included agricultural elements

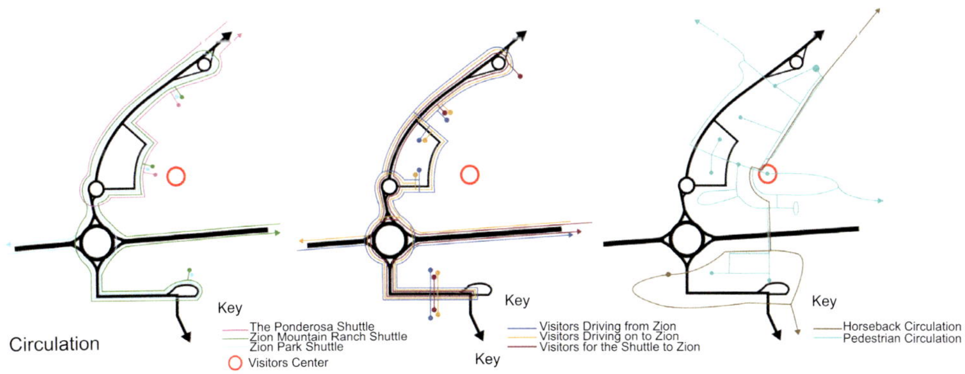

Fig. 10.7. Several of the student design proposals incorporated a roundabout on State Route 9 to access the proposed visitor contact centre, shuttle parking and other visitor amenities (Evan Tanasiuk, Utah State University LAEP, 2018).

into the designed experience for Zion's eastern gateway, which provide opportunities for visitor experiences at the east entrance that are unavailable at the park's western gateway (Table 10.1). Moreover, the east entrance design not only integrates the cultural and historical aspects of land use in the area, it also offers opportunities for local farmers and ranchers, food processors, restaurants and specialty stores to participate in park tourism (Wilson *et al.*, 2006), thus reducing economic leakages and enhancing community welfare. National park visitors are often cultural tourists (Slocum and Curtis, 2016) and thus are highly interested in experiencing the historical and cultural aspects of an area, including local foods.

The students' work provided landowners a new perspective on the planning and design of the east Zion entrance just as efforts to bring the project to reality were accelerating. Engaging the landowner in the students' design process created a synergistic creative environment wherein design solutions were explored collaboratively. The realizations shared by both the students and the landowners, and the visual resources produced by the class, guided the landowner discussions with adjacent public land managers, as well as informing the intent and thematic design of the project by private consultants.

10.4.1 Project developments post plan

The proposed developments for Zion National Park's eastern gateway will shortly be reality. In 2021, the Zion National Park Forever Project announced plans to construct the new visitor centre in collaboration with ZMR, Zion National Park, the Zion Forever Project, BLM and Kane County. The 5000 square foot facility, located on land currently owned by ZMR at the intersection of State Route 9 and North Fork Road, will provide backcountry permits, retail, educational and interpretive programming, and host events. Constructed of sustainably and locally sourced materials, the facility will provide agricultural-based visitor experiences, including several key programmatic design elements proposed by both professional practitioners and USU students (Figs 10.8–10.10). They include heritage farming activities, a farmers' market, orchards, and connectivity to explore the surrounding agricultural landscape on private lands, which will be protected by conservation easements (Zion National Park Forever Project, 2021).

Table 10.1. Agricultural and food tourism activities.

Agritourism	Farm stays	Farm-based activities	Pick-your-own produce	Farm stands	Fishing and hunting
Buildings/ structures	Food processing facilities	Wineries, breweries, distilleries	Farmers' markets	Food stores	Food-related museums
Consumer shows/ festivals	Food and drink shows	Cooking demonstrations	Food festivals	Beer/wine festivals	Harvest festival
Consumption	Dining at restaurants	Picnics utilizing locally-grown foods	Local sourcing	Tasting local recipes	Food-based souvenirs
Touring	Wine regions or routes	Agricultural regions	City food districts	Food routes	Gourmet trails
Land uses	Farms	Orchards	Vineyards	Restaurant districts	
Educational	Cooking schools	Wine or beer tasting	Wine and beer making or blending	Food and beverage magazines	Food and wine/ beer pairings

Adapted from Everett and Slocum (2013)

Fig. 10.8. Student design proposals included agricultural-themed design elements for the proposed visitor centre and related structures (Ethan Balls, Utah State University LAEP, 2018).

In addition, the site will include several of the other design features proposed by USU students and professional firms, including an outdoor amphitheatre for educational programming hosted by the National Park Service Junior Ranger Program, a natural spring water filling station, and the parking, restrooms and other ancillary infrastructure necessary for shuttle connectivity to Zion National Park and destinations throughout Kane County's regional communities. The NPS and Zion Forever Project are anticipated to enter into a long-term lease to use the facility (Zion National Park Forever Project, 2021).

According to Kane County Commissioner, Brent Chamberlain, ZMR is donating 18 acres of private land near Zion National Park's eastern gateway to Kane County, which will use a $15 million Community Impact Board (CIB) loan to fund construction of the new visitor centre. The visitor centre will be staffed by personnel from Zion National Park, Zion Forever and Kane County. According to Chamberlain, Kane County is also establishing a regional transportation authority (RTA) to connect the City of Kanab, as well as other communities in the region, to the new visitor centre, which will include a transit station, parking and access to approximately 40 miles of new trails near the park boundary (*Fox 13 News*, 2021, March 26), funded, in part, by a $150,000 grant from the Utah Office of Outdoor Recreation (Shoup, 2019). The trail network will accommodate bicycling, hiking and horseback riding on private property, connecting to BLM lands north and south of ZMR. Future planning calls for connecting trails to existing trails located within Zion National Park.

According to Zion National Park Forever Project, the new visitor centre project and its associated amenities 'will serve as a model for public–private partnership and will extend the visitor experience across the greater Zion landscape' (Zion National Park Forever Project, 2021, para. 5). To support the goals outlined in the regional transit study, Zion National Park recently received a $33 million federal transit grant to replace its shuttle fleet with zero emission, battery electric buses (NPS, 2021a). Economic analyses of the proposed east gateway

Fig. 10.9. Student design proposals integrated orchards, vegetable gardens and local food-based activities into the visitor experience throughout the site (Ethan Balls, Utah State University LAEP, 2018).

development estimates the creation of 451 new jobs, and the development should result in $29.6 million additional annual spending by park visitors (Leaver and Pace, 2021).

10.5 Significance, Implications and Opportunities for Future Inquiry

While the most visited national parks usually have well developed gateway communities, this project represents a potential model for similarly situated communities and areas, particularly:

- areas near national parks that have been previously undeveloped due to natural barriers, or large privately-owned parcels under natural resource extractive or agricultural land uses;
- less developed areas near large public landholdings that are managed by other public agencies;
- areas where problems cannot be solved unilaterally by a single agency, government agency or private consortium;
- situations in which PPP is key to addressing planning and design challenges in a thoughtful and effective way; and
- locations where local knowledge is key to connecting public land managers with landowners.

Through a pro-local approach, some visitation impacts can be externalized in an intentional

Fig. 10.10. Students used on-site sketching as well as digitally produced drawings to convey their vision for how the site might enhance visitor experiences (Evan Tanasiuk, Utah State University LAEP, 2018).

and thoughtful way. Time will tell if this PPP model is an effective approach for gateway community development outside of national parks. As design proposals progress from concept to reality, this alternative approach to the creation of a gateway for Zion National Park may demonstrate that strategic, partnership-focused, master planning and design can proactively address some of the key challenges facing gateway communities. Alternatively, this development may find that Zion's new gateway community experiences similar issues to those presently occurring on the park's southwest entrance.

10.5.1 Challenges for PPPs

Land ownership is dispersed among many different owners with disparate interests, values and visions for the future, which can often pose a challenge for effective PPPs (Farris, 2001). In these instances, a development corporation may be needed to facilitate assembly of individual parcels into land sufficient to support development that can facilitate visitor experiences anticipated in a national park environment. While urban areas have seen a renaissance of farmers' markets and concentrated urban agriculture projects, rural landscapes have much more potential to support a broader range of agricultural and food-based activities. However, in situations with a limited number of stakeholders, a single stakeholder may also hold power to prevent development if their interests are not satisfied. Additionally, where natural features, including topography, hydrological features and sensitive ecosystems limit the range of potential activities suited for the site, partnerships play an important role in allowing planning and design to address these challenges at an adequate scale.

Recent rural migration has accelerated the discovery of, and development within, rural western landscapes. Less developed places not located near conventional national parks have received attention and visitation by people seeking natural amenities and outdoor recreation, as well as cultural and historical experiences (Everett and Slocum, 2013; Slocum and Curtis, 2016). Consequently, locations where rapid growth and visitation pressures are manifested have begun to spill over beyond those places traditionally viewed as gateway communities and into communities on the fringes of the public–private land interface to encompass places that have strong identities rooted in their working agricultural landscapes (Howe *et al.*, 2012; Stoker *et al.*, 2021). When visitor-driven infrastructure is planned and constructed in these areas, there are strong opportunities for integrating authentic agricultural and food-based experiences into the visitor-focused infrastructure. PPPs may have greater potential for success where land ownership is concentrated in the hands of fewer landowners, in larger parcels, and in land that has fewer natural constraints for broader development potential. Often, agricultural lands possess these qualities.

10.5.2 Opportunities for future inquiry

Complex problems of cross-boundary transportation, infrastructure and park visitor experiences cannot be effectively solved within the bounds of statutorily and geographically constrained public lands. Nor do private interests hold the ability to effectively absorb significant tourism and development impacts without coordination with federal, state and local government agencies. In gateway communities, effective coordination requires a willingness to engage in conversation, find common ground among various groups with different interests, and successfully negotiate mutually beneficial solutions. When conceived as design problems, conversations on transforming existing circumstances into preferred conditions (Simon, 1988) can be catalysed and enhanced through the persuasive power of an image (Arnheim, 1969; Barry, 1997). Professional design firms are fully capable of conceiving, and illustrating, alternative spatial solutions for community-based design problems. However, design students have particular licence to propose unconventional solutions that can broaden the range of possibilities considered. Within traditionally-minded agricultural communities in the western US, land-grant universities, due to their credibility from assisting local communities and rural areas, have a unique opportunity to engage with and mediate between different or conflicting interests. In the case of the east Zion gateway, these images were created by students in a land-grant-based design programme. However, addressing other dimensions of complex project design problems, including their economic, social and environmental aspects, presents opportunities for other land-grant-based programmes to engage in gateway development.

Designing a new gateway community for a national park is a once in a lifetime opportunity. This particular development project is a work in progress. As Zion's eastern gateway develops, its processes and outcomes should be analysed for additional insights. Future studies may look at exploring relationships between nationally designated heritage areas and agricultural tourism activities, and externalization of national parks' visitation impacts on adjacent gateway communities and the surrounding regional landscape. While this study describes the application of a PPP in the creation of a new gateway community for one of the most visited national parks in the US, future studies might explore how PPPs are applied within other typologies of gateway communities, including those situated near other non-NPS agency-managed public lands, such as those of the US Forest Service and the BLM, that are managed for multiple uses.

References

Allen, S. (2012) President's message. *Ranger: The Journal of the Association of National Parks* 28(3), 1.

Arnheim, R. (1969) *Visual Thinking*. University of California Press, Berkeley, CA, USA.

Barry, A.M. (1997) *Visual Intelligence: Perception, Image, and Manipulation in Visual Communication*. State University of New York Press, Albany, NY, USA.

Bohdanowicz, P. (2006) Environmental awareness and initiatives in the Swedish and Polish hotel industries – survey results. *International Journal of Hospitality Management* 25(4), 662–682. DOI: 10.1016/j.ijhm.2005.06.006.

Chin, J.T. (2021) The shifting role of public–private partnerships in vacant property redevelopment. *Land Use Policy* 105(3), 105430. DOI: 10.1016/j.landusepol.2021.105430.

Daigle, J.J. (2008) Transportation research needs in national parks: a summary and exploration of future trends. *The George Wright Forum* 25(1), 57–64.

Deller, S.C., Tsai, T.H., Marcouiller, D.W. and English, D.B. (2001) The role of amenities and quality of life in rural economic growth. *American Journal of Agricultural Economics* 83(2), 352–365. DOI: 10.1111/0002-9092.00161.

Department of the Interior, Bureau of Land Management, and Office of the Solicitor (1976) The Federal Land Policy and Management Act, as amended. U.S. Department of the Interior, Bureau of Land Management Office of Public Affairs, Washington, D.C., p. 69.

Everett, S. and Slocum, S.L. (2013) Food and tourism: an effective partnership? A UK-based review. *Journal of Sustainable Tourism* 21(6), 789–809. DOI: 10.1080/09669582.2012.741601.

Farris, J.T. (2001) The barriers to using urban infill development to achieve smart growth. *Housing Policy Debate* 12(1), 1–30. DOI: 10.1080/10511482.2001.9521395.

Finnessey, L. (2012) The Negative Effects of Tourism on National Parks in the United States. Honors theses, Johnson & Wales University - Providence Campus. 4. Available at: https://scholarsarchive.jwu.edu/student_scholarship/4 (accessed 7 February 2022).

Fox 13 News (2021) In-Depth: *Possible expansion of east entrance to Zion National Park* (2021, March 26).

Frauman, E. and Banks, S. (2011) Gateway community resident perceptions of tourism development: incorporating importance-performance analysis into a limits of acceptable change framework. *Tourism Management* 32(1), 128–140. DOI: 10.1016/j.tourman.2010.01.013.

Gilroy, L., Kenny, H. and Morris, J. (2013) Parks 2.0: operating state parks through public-private partnerships. *Reason Foundation Policy Study* 419.

Gimmi, U., Schmidt, S.L., Hawbaker, T.J., Alcántara, C., Gafvert, U. *et al.* (2011) Increasing development in the surroundings of U.S. National Park Service holdings jeopardizes park effectiveness. *Journal of Environmental Management* 92(1), 229–239. DOI: 10.1016/j.jenvman.2010.09.006.

Grimsey, D. and Lewis, M.K. (2002) Evaluating the risks of public private partnerships for infrastructure projects. *International Journal of Project Management* 20(2), 107–118. DOI: 10.1016/S0263-7863(00)00040-5.

Hamin, E.M. (2001) The US National Park Service's partnership parks: collaborative responses to middle landscapes. *Land Use Policy* 18(2), 123–135. DOI: 10.1016/S0264-8377(01)00006-0.

Howe, J., McMahon, E.T. and Propst, L. (2012) *Balancing Nature and Commerce in Gateway Communities*. Island Press, Washington, DC, USA.

IRMA Portal (2021) *Traffic Counts at East Entrance.* Integrated Resource Management Applications. Available at: https://irma.nps.gov/STATS/SSRSReports/Park%20Specific%20Reports/Traffic%20Counts?Park=ZION (accessed 7 February 2022).

Kane County, General Plan (2018, August) *For the Physical Development of the Unincorporated Area Pursuant to Section 17-27a-403 of Utah State Code*. Available at: https://kane.utah.gov/gov/dept/planning/draft-of-2018-kane-county-general-plan/ (accessed 7 February 2022).

Kasim, A. (2006) The need for business environmental and social responsibility in the tourism industry. *International Journal of Hospitality & Tourism Administration* 7(1), 1–22. DOI: 10.1300/J149v07n01_01.

Kwak, Y.H., Chih, Y. and Ibbs, C.W. (2009) Towards a comprehensive understanding of public private partnerships for infrastructure development. *California Management Review* 51(2), 51–78. DOI: 10.2307/41166480.

Leaver, J. and Pace, L. (2021, February) *An Economic Analysis of Zion National Park Scenarios*. University of Utah Kem C. Gardner Policy Institute, Salt Lake City, UT, USA. Available at: https://gardner.utah.edu/wp-content/uploads/ZionsReport-Feb2021.pdf (accessed 7 February 2022).

National Park Service (n.d.) *National Park Friends Alliance*. National Park Foundation, Washington, DC, USA. Available at: https://www.nationalparks.org/about-foundation/partnerships/community-partnerships/national-park-friends-alliance (accessed 7 February 2022).

National Park Service (2006) *Management and Policies*. Available at: https://www.nps.gov/policy/mp2006.pdf (accessed 26 August 2016).

National Park Service (2016) *Frequently Asked Questions*. Available at: https://www.nps.gov/aboutus/faqs.htm (accessed 30 May 2016).

National Park Service (2017) *Cultural Landscapes*. Available at: https://www.nps.gov/subjects/culturallandscapes/sustainability.htm (accessed 17 May 2021).

National Park Service (2021a) *Zion National Park Receives $33 Million Federal Transit Grant*. Available at: https://www.nps.gov/zion/learn/news/zion-national-park-receives-33-million-federal-transit-grant.htm (accessed 7 February 2022).

National Park Service (2021b) *Visitation Numbers*. Available at: https://www.nps.gov/aboutus/visitation-numbers.htm (accessed 7 February 2022).

Sealey, K.S. and Smith, J. (2014) Recycling for small island tourism developments: food waste composting at Sandals Emerald Bay, Exuma, Bahamas. *Resources, Conservation and Recycling* 92(1), 25–37. DOI: 10.1016/j.resconrec.2014.08.008.

Shoup, M. (2019) Gateway community' planned for east entrance to Zion National Park (2019, May 14). *St. George News*. Available at: https://www.stgeorgeutah.com/news/archive/2019/05/14/mks-gateway-community-planned-for-east-entrance-to-zion-national-park/#.YIYyqi2ZOL8 (accessed 7 February 2022).

Simon, H.A. (1988) The science of design: creating the artificial. *Design Issues* 4(1/2), 67–82. DOI: 10.2307/1511391.

Sleipness, O., Powell, J., Anderson, D., Evans, D., McCann, R. *et al.* (2019) Extension-based community engagement project contributions to landscape architecture core competencies and professional values. *NACTA Journal* 63(2), 37–46.

Sleipness, O., Licon, C., Powell, J., Anderson, D., Evans, D. *et al.* (2020) Designing a legacy: Contributions of university design engagement toward Zion's iconic western image. *Landscape Research Record* 9, 94–109.

Slocum, S.L. (2017) Operationalising both sustainability and neo-liberalism in protected areas: implications from the USA's National Park Service's evolving experiences and challenges. *Journal of Sustainable Tourism* 25(12), 1848–1864. DOI: 10.1080/09669582.2016.1260574.

Slocum, S.L. and Curtis, K. (2016) Assessing sustainable food behaviour of national park visitors: Domestic/on vacation linkages, and their implications for park policies. *Journal of Sustainable Tourism* 24(1), 153–167.

Stoker, P., Rumore, D., Romaniello, L. and Levine, Z. (2021) Planning and development challenges in western gateway communities. *Journal of the American Planning Association* 87(1), 21–33. DOI: 10.1080/01944363.2020.1791728.

Tang, L., Shen, Q. and Cheng, E.W. (2010) A review of studies on public–private partnership projects in the construction industry. *International Journal of Project Management* 28(7), 683–694. DOI: 10.1016/j.ijproman.2009.11.009.

Thering, S. and Chanse, V. (2011) The scholarship of transdisciplinary action research: toward a new paradigm for the planning and design professions. *Landscape Journal* 30(1), 6–18. DOI: 10.3368/lj.30.1.6.

Thomas, C. and Koontz, L. (2016) 2015 National Park Visitor Spending Effects: Economic Contributions to Local Communities, States, and the Nation. Natural Resource Report NPS/NRSS/EQD/NRR – 2016/1200. National Park Service, Fort Collins, CO, USA.

UDOT (2013) *R930. Transportation, Preconstruction. R930-6. Access. Management*. Utah Department of Transportation, Salt Lake City, UT, USA. Available at: https://www.udot.utah.gov/main/uconowner.gf?n=11066229893635233 (accessed 7 February 2022).

Utah State University LAEP (2018) *Zion Mountain Ranch: The Junction Design*. Landscape Architecture and Environmental Planning, Logan, UT, USA.

Vogel, S. (2013) National parks to offer healthier food under new standards. *The Washington Post*, June 9, 2013. Available at: https://www.washingtonpost.com/politics/national-parks-to-offer-healthier-food-under-new-standards/2013/06/09/64d9cd44-cd8a-11e2-8845-d970ccb04497_story.html (accessed 2 June 2016).

Weber, J. and Sultana, S. (2013) The Civil Rights movement and the future of the national park system in a racially diverse America. *Tourism Geographies* 15(3), 444–469. DOI: 10.1080/14616688.2012.675515.

Wilcox, C. (2015) Is ecotourism an Oxymoron? A new study suggests these "green" vacations might hurt wildlife. *The Washington Post*, 10 October 2015. Available at: https://www.washingtonpost.com/news/speaking-of-science/wp/2015/10/10/is-ecotourism-an-oxymoron-a-new-study-suggests-these-green-vacations-might-hurt-wildlife/ (accessed 10 September 2015).

Wilson, J.B., Thilmany, D. and Watson, P. (2006) The role of agriculture in western states: Place-specific and policy factors influencing recreational income for producers. *The Review of Regional Studies* 36(3), 381–399.

Winks, R.W. (1996) The National Park Service act of 1916: a contradictory mandate. *Denver University Law Review* 74, 575.

Zion National Park Forever Project (2021, April 25) *Establishing a visitor contact station for Zion's east entrance*. Available at: https://zionpark.org/projects/establishing-a-visitor-contact-station-for-zions-east-entrance/ (accessed 7 February 2022).

11 Stories to Engage Protected Area Neighbouring Communities in Tourism Planning and Practice

Gianna Moscardo*
College of Business, Law and Governance, James Cook University, Australia

Abstract
A major theme in the broader literature on community development planning is the need to be more flexible and inventive, calling for greater creativity in the design of tourist experiences. These calls are especially applicable to gateway communities that rely upon, and are impacted by, tourism. The 'story turn' in social sciences offers an opportunity to pursue this creativity. This chapter uses Magnetic Island (MI), a gateway community to two protected areas (PAs), as a case study to explore how new stories of tourism destinations might enhance the sustainability and resilience of tourism. A simple story-based model of tourism is presented, and the role of stories in two aspects of this system is identified: the co-creation of new stories of tourism in the interactions between tourism planners and destination residents; and the use of stories in tourist experience design. Possibilities are presented to demonstrate the value of this story-based approach.

Keywords: community engagement, tourism planning, gateway communities, stories, experience co-creation

11.1 Introduction

Living within, next to, or at the gateway to a protected area (PA) is a double-edged sword. On the one hand, communities in these locations have access to areas of outstanding natural beauty and/or cultural significance, as well as opportunities to benefit from the development potential of economic activities, especially tourism, associated with PAs (Snyman and Bricker, 2019). On the other hand, they can also bear the brunt of the negative impacts of these economic activities and are often excluded from the governance of these PAs (Stoker *et al.*, 2020). Magnetic Island (MI), located 8 km off the northeastern coast of Australia, is one of these communities. More than half of this 52 km^2 tropical island is a national park, and the whole of the island lies within the boundaries of the Great Barrier Reef World Heritage Area and Marine Park (Queensland Parks & Forest, 2020). It is both a gateway to these two PAs and home to approximately 2400 residents. MI is accessible by passenger-only ferry (20 minutes) or car ferry (40 minutes), leaving from the regional city of Townsville. This chapter will use this specific community as a case study to examine the potential role of stories for engaging communities in both tourism planning and practice.

The chapter presents a story-based conceptual framework for tourism, identifying a number of different types of stories and roles of stories in

*gianna.moscardo@jcu.edu.au

DOI: 10.1079/9781789249033.0011

the tourism system. It focuses on two of these roles. One is the use of stories in destination governance as a tool for encouraging more extensive and innovative resident engagement in tourism planning. This section draws upon recent literature from the general community development planning area focused on stories and creativity as tools for encouraging more widespread and innovative resident engagement in community governance and planning. The second role is the use of stories to structure and organize tourist experience opportunities that offer more direct interaction and co-creation between tourists and residents. To date, much of the discussion of stories in tourism in general, and in the design of tourist experiences, has been focused on tour guide staff telling specific destination stories to tourists. The present discussion moves beyond storytelling to story co-creation and focuses on how this co-creation can offer community members greater opportunities to directly engage in tourism, and with tourists, to influence the nature of tourism, both within their communities as well as within the adjacent PAs.

The chapter offers two alternative approaches to the challenges that gateway communities face. The first is the use of more creative techniques to engage community residents in planning processes. While this has been suggested and trialled in recent years in the broader planning literature, in tourism planning these techniques have not been adopted, or even much mentioned (Moscardo, 2019a). The second is the use of stories as a framework for developing tourist experience opportunities that allow for the co-creation of new stories between residents and tourists (Moscardo, 2020a). After briefly reviewing the case study setting, the chapter will define stories and describe their different roles in the tourism system. It will then examine, in more detail, the role of stories in tourism planning and then in the design of tourism experience opportunities that better link local residents in gateway communities to tourists and tourism business opportunities.

11.2 Tourism and Magnetic Island

MI has a long and troubled tourism development history with 20 years of community conflict over a proposed large-scale tourism resort development that was ultimately not completed (Moscardo and Murphy, 2013). The conflict resulted in considerable environmental and social damage and a general lack of trust in both government agencies and external tourism businesses (Moscardo and Murphy, 2015). For many years, the dominant story of tourism among MI residents was that tourism was an external destroyer of environments and local lifestyles (Moyle *et al.*, 2010). In the last decade, the community, in partnership with local tourism businesses, local government and the relevant PA management agencies, has sought to create a new story of tourism that is focused on improving the overall sustainability of the community and finding innovative ways to better manage tourism. This new story is one of MI as a sustainable suburb focusing on being a gateway to both PAs, and it has created a resurgence of local enthusiasm for tourism.

The author has been involved in several tourism planning and marketing activities in this community over the last decade. Consistent with the guidelines for case study analyses (Yin, 2012; De Massis and Kotlar, 2014), this case study uses results from these planning and marketing activities, which included visitor surveys and structured group workshop activities focused on tourism futures scenario building, as well as from archival analyses of tourism development plans and reports, news media and stories posted online by MI residents, visitors and professional travel bloggers. The aim is to use the case study to outline how stories could be used to more effectively engage community residents in tourism planning and practice.

11.3 A Story-based Framework for Tourism

Even a casual examination of current practitioner discussions of tourism marketing reveals the existence of a story-turn reflecting widespread research into stories in marketing (Moscardo, 2020b). This work in marketing, in turn, recognizes the existence of a story-turn in other social science disciplines, including anthropology (Dunn, 2017), psychology (Popova, 2015) and sociology (Polletta *et al.*, 2011). Tourism academics have been much slower to examine stories, with most work emerging from studies of user-generated content, such as travel blogs and reviews (Volo, 2010). In these recent discussions of tourist writings, stories are often confused with narratives, and the research is rarely informed by existing social science research into stories, making it difficult to see any coherent understanding of the importance of stories emerging. The wider social science story-turn does, however, offer a clear definition of a story and how it differs from the concept of a narrative. Based on this work, this chapter will describe narratives as the broader category of all types of descriptions of events told in some sort of temporal order (Polletta *et al.*, 2011). A story is defined as a specific type of narrative with the following features:

- the aim is to elicit emotional responses and entertain the audience;
- it must include a character or set of characters who must be confronted by an event, incident or challenge and make a decision in response to this event;
- it reveals the responses of these characters to each other and the event, as well as the outcomes of the decisions made; and
- the events must be causally related to each other and to the motives and personalities of the characters (Adaval and Wyer Jr, 1998; Popova, 2015; Moscardo, 2018).

Stories have been shown to play critical roles in memory (Goldstein, 2015), learning (Popova, 2015), persuasion (Kent, 2015), personal and social identity (Fivush, 2011), the development and maintenance of social relationships (Laszlo, 2008), and the transmission and maintenance of culture (Dunbar *et al.*, 2016). Given the importance of stories in human memory, understanding and action, it is not surprising to find that marketers have moved towards greater use of storytelling in brand development and product promotion (Woodside, 2010). While destination promotion stories have become important in tourism marketing (Stevenson, 2019), stories can also be linked to other aspects of the tourism system. The simple model in Fig. 11.1 includes the three main components of the tourism system: the tourists; tourism practitioners, including both public and private sector organizations involved in tourism; and destination communities.

Stories exist in the three main intersections identified in the tourism system. The first is the intersection between tourism practitioners and tourists, and in this space, stories are mostly told to tourists as part of destination or tourism business promotion. These storytellers can be formal destination marketing organizations (DMOs), as well as previous tourists and social media influencers. Although the use of stories in destination promotion was increasing rapidly pre-Covid, many destinations were still using more traditional approaches of listing attractions and generic features. MI is one of these latter destinations. A simple content analysis of formal destination promotional material, including professional travel writers and social media influencers, reveals almost no use of stories in presentations of MI. Instead, promotional material highlights beaches, weather, wildlife, relaxation and access to nature. This suggests potential for MI to further develop its marketing by developing and promoting an MI 'storyworld'. A storyworld refers to a set of stories linked by a place, event or central character. Table 11.1 summarizes some of the existing stories that could be used to build an MI storyworld.

The second role identified in Fig. 11.1 exists in the interactions between tourism practitioners, especially destination planners and developers, and destination communities, especially residents. In this interaction, the stories are about tourism itself, rather than stories about the destination. The third role for stories exists in the area of tourist experiences. Here the stories are more mixed, including stories of the destination, stories of the residents, stories of the tourism businesses and staff and stories of the tourists themselves. In both these roles, it is important to distinguish between storytelling and story creating. In storytelling, a single person or organization controls and presents a story to a largely passive audience. In story creation, multiple people can interact and cooperate to use the destination setting and available activities to create a new, and often personally relevant, story (Moscardo, 2017). This distinction is an important one as we explore these two roles in more detail in the following sections.

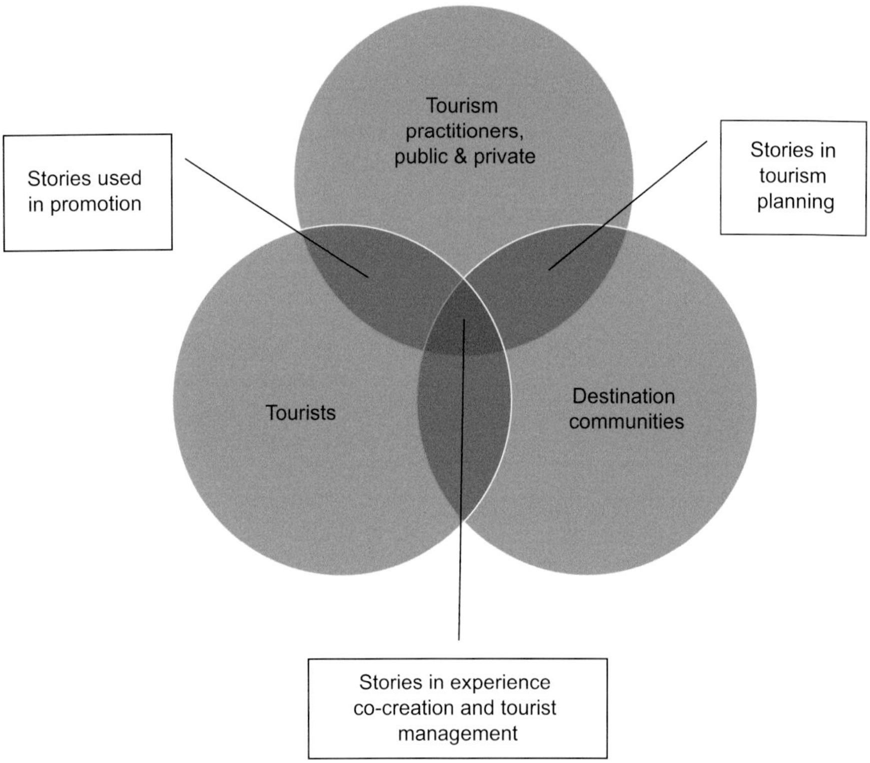

Fig. 11.1. Stories in the tourism system (adapted from Moscardo, 2020b).

11.3.1 The role of stories in tourism planning

There are several dimensions to the use of stories in tourism planning – the stories of tourism, in general, told to destination communities by tourism developers and planners, the stories about the nature of tourism at the specific destination that can emerge from the interactions between the various stakeholders in tourism, and the use of stories and storytelling in planning processes. Saarinen *et al.* (2017) present six approaches to tourism planning that can be matched to three core stories of tourism, all of which have tourism as the chief protagonist. The first two planning traditions, boosterism and economic planning, are based on promoting tourism growth, with little or no questioning of its impacts or understanding of how this growth actually benefits the destination (Hall, 2008). In these planning approaches, tourism is presented as a saviour, or hero, like Gandalf in Tolkien's *The Lord of the Rings*. Spatial, environmental, and community-based planning is then developed in response to recognition of the extensive negative impacts that uncontrolled or unmanaged tourism brings to many destinations (Hall, 2008). In these traditions, tourism is a destroyer or villain, like Tolkien's Sauron. Sustainable and integrated planning attempts to balance these two competing visions with a story of tourism as a powerful force with potential for good, but only if it can be controlled (Hall, 2008), not unlike 'the Ring' in both of Tolkien's epic travel sagas. Saarinen *et al.* (2017) argue for a seventh tradition called new public planning, in which governments adopt the same planning processes as used in businesses.

In the years prior to the Covid-19 pandemic, tourism as a destroyer emerged strongly in public discussions of overtourism (Koens *et al.*, 2018). For some, the critical role of international tourism in the spread of the virus is further evidence of its destructive character. Discussions of

Table 11.1. Possible inclusions in an MI storyworld.

Type of story	Specific examples
Human history	Indigenous stories Captain Cook and his naming of the destination based on a false belief that it had some magnetic powers and was interfering with his instruments World War II history as a fort and communications post (see https://parks.des.qld.gov.au/parks/magnetic-island/about/culture for more details) Earlier European colonial use as a quarantine station and leisure location including history of specific sites such as the Picnic Bay Jetty or Centaur House (see https://www.slq.qld.gov.au/blog/centaur-house-memorial-hostel-magnetic-island) The conflict over Magnetic Quays Specific characters who have lived on the island The stories of the various shipwrecks around MI (see https://www.abc.net.au/news/2018-05-31/forgotten-shipwrecks-of-magnetic-island-tourism-bid/9795918)
Natural history	Wildlife and unique flora Connection to the Great Barrier Reef
Current lifestyles	The push to be a sustainable suburb Stories of local tourism/business owners Netflix series focused on Izzy Bee – *The Koala Whisperer* (see https://inqld.com.au/culture/2020/09/24/izzys-magnetic-charm-with-koalas-brings-queensland-to-worldwide-audience/)
Fiction wholly or partly set on MI	*The Sister's Keeper* (2020), B. Hannay *Magnetic Island: A Novel* (2019), H. Cummins *A Tango with the Dragon* (2015), G. Goldsmith *Tiwi: the Lost Baby Koala of Magnetic Island* (2008), J. Quintana *Affection* (2007), I. Townsend *Girl with a Monkey* (1958), T. Astley
Stories told by tourists	In travel diaries, blogs and reviews (see https://www.travelblog.org/Oceania/Australia/Queensland/Magnetic-Island/blog-804267.html)

tourism during the pandemic have been characterized as a war between those who see tourism as a saviour that must be restored as soon as possible using any means available, and those who see tourism as a powerful force in need of control (Higgins-Desbiolles, 2020). For the former, the Covid-19 virus is a monster blocking the pathway of tourism progress, while for the latter the virus is a prophet sounding a warning about the need to take control of tourism. Grimwood *et al.* (2019) present similar plots in their discussions of stories told by Indigenous groups about colonial settlers, and how current tourism development often builds upon these settler stories. These authors go further to argue that the co-creation of new stories by Indigenous groups offers a way to disrupt these disempowering tourism development processes.

The stories of tourism discussed in these sections share a common feature, in that they are written and presented by tourism marketers, developers or planners and told to destination communities. When tourism planners and/or developers present a simple story of tourism as a saviour to a largely passive audience of destination residents, then it is difficult to generate the type of community participation required to support sustainable outcomes. An alternative approach to tourism planning and development is co-creating new stories of tourism in which the destination community is both a core creator of the tourism story and the chief protagonist, learning how to use and control tourism as a tool to achieve their wellbeing needs and aspirations.

As previously noted, discussions of tourism on MI have been dominated for nearly 30 years by a deep and bitter conflict over a single large-scale resort development. The Magnetic Quays project was a classic example of tourism as an economic hero, a story told to the local community by a group of external business investors (see Moscardo *et al.*, 2014 for more detail). In this case, considerable opposition to the proposal was generated by members of the MI community because of concerns about damage to the protected environments, especially the Great Barrier Reef, which are important aspects of resident quality of life. The level of conflict eventually led to the proposal being shelved with some significantly smaller developments completed. In response to the conflict, regional and state DMOs began to recognize the need to listen to resident concerns and aspirations and to reposition tourism development plans to reflect resident interest in becoming a sustainable suburb. The Tourism Opportunity Plan (Tourism Queensland, 2009, p. 19), for example, sought to support 'a range of initiatives to further develop Magnetic Island as a benchmark sustainable community and tourist destination. Magnetic Island offers a unique opportunity to showcase the application of a wide range of sustainable initiatives and technologies in a real community'. This overall goal was based on a new story of MI tourism, one where the island, and its desire to be a sustainable community, was the main protagonist, not tourism.

As part of this alternative destination community wellbeing approach to tourism planning (see Moscardo and Murphy, 2013), the author was involved in a series of workshops conducted with local MI residents in which this new story of tourism was developed further by asking residents to create more specific stories of what MI tourism might be like in the future. These stories are typically called futures scenarios and are a core element of futures thinking (Yeoman and Postma, 2014; Burnam-Fink, 2015). It has been argued that the process of creating a story, or scenario, about the future helps people to move beyond their 'traditional operational and conceptual comfort zone' (Amer *et al.*, 2013, p. 25). Futures scenarios are commonly created by 'experts' and presented to the target audience for comment (Oteros-Rozas *et al.*, 2015). The key to the present case was to take a participatory approach and encourage MI residents to create their own aspirational stories of tourism in the future. These stories shared several common themes:

- the importance of the natural environment and the two PAs for the wellbeing of the MI community;
- the importance of sustainable lifestyles, based on renewable energy, lower carbon emissions, water conservation programmes and sustainable building design that allowed for continued appreciation of these protected natural areas (PNAs); and
- the need to develop small scale, low impact, locally owned tourism businesses based around providing positive intensive tourist experiences in these natural environments.

As a consequence of this new story-based tourism planning approach, efforts were directed towards the development of educational tourism, interpretive walking and snorkelling trails, health and wellbeing-based tourism experiences, and more extensive and overt use of sustainability practices across the tourism sector. It is worth noting that this approach has proven to be an effective strategy for building a resilient, as well as sustainable, destination. As these strategies enhanced the perception of MI as a positive destination for local wellbeing, MI tourism has generally recovered reasonably well following Covid-19 shutdowns (Queensland Times, 2020).

The use of stories to encourage community engagement in tourism planning is not restricted to the development and analysis of participatory futures scenarios. Stories have been identified as effective tools in developing shared understandings across cultural boundaries, making various forms of storytelling important tools in encouraging open dialogue between different groups involved in tourism planning and practice (Stevenson, 2019). A major shift in planning, in general, has been towards the use of more flexible and creative approaches to engage community members in planning processes (Moscardo, 2019a). Such processes include stories of tourism presented to communities through the use of art, traditional storytelling techniques and/or theatre (Da Costa, 2013; Amy, 2016; Cowey, 2017).

11.3.2 The role of stories in experience co-creation

A lack of community engagement in tourism planning processes creates two other major barriers to tourism generating net positive benefits for destination resident wellbeing: conflict over tourism development proposals and a lack of destination resident participation in tourism businesses and tourist activities (Moscardo, 2011; Hewlett and Edwards, 2013; Ruhanen, 2013; Nunkoo, 2015). A lack of direct involvement by destination residents in tourism businesses and tourist activities creates problems, such as economic leakage, perceptions of tourists as unwelcome outsiders, conflict between residents and tourists using the same spaces and facilities and a lack of direct benefits for residents (Moscardo, 2019b). An alternative to this situation is to use stories as the basis for the creation of tourist experiences that bring together destination residents and tourists. These experiences, then, offer opportunities for residents to benefit from tourist activities through: employment; the provision of business services to support these experiences; celebrating the positive aspects of their destination; enhancing a sense of community pride by extending leisure, culture, and recreational opportunities and facilities available to residents; and/or by providing opportunities to build bridging social capital. There are two ways stories can be used in tourist experience design: the telling of destination stories to tourists and the provision of opportunities for tourists to create their own personal stories (Moscardo, 2020a).

The telling of destination stories to tourists has long been a major part of tourist experiences. Table 11.2 summarizes the main methods used to tell destination stories, with examples from a range of destinations and existing, proposed and potential examples for MI. The stories told can be nonfiction accounts of the history and events of the destination, as are often told by traditional tour guides, or fictional stories used to highlight features of the destination and its history and culture (for example, the Historium in Bruges, Belgium, https://www.historium.be/en/discover-historium/vr). The storyteller can be a third-person narrator telling the story from an external perspective (for example, Casanova-themed guided tours of Venice, https://www.tours-italy.com/tour/private-venice-life-times-of-casanova-walking-tour_125), or can act as a first-person character telling the story from the perspective of a protagonist in the story (for example, the Night Watchman tours in Rothenburg ob der Tauber, Germany, https://www.rothenburg-tourismus.de/en/plan-book/top10-activities/night-watchman). The main protagonists in the story can be human or not. Animals can be central characters in stories told about the natural history of a destination, such as the story of Kimberley, a virtual tree-kangaroo, presented at the Malanda Falls Interpretive Centre in the Wet Tropics Rainforest World Heritage area (https://www.tree-kangaroo.net/events/upcoming-events-official-launch-kimberley-virtual-reality-tree-kangaroo-experience), as can geographical features, such as rivers and volcanoes. For example, the Mississippi River features as a central character in stories told in the Delta Cultural Centre in Depot, Arkansas, USA (https://www.deltaculturalcenter.com/venues/depot).

Another outcome of the MI resident tourism futures workshops was acknowledgement of Indigenous history on MI and suggestions to include Traditional Owners more in tourism. There is evidence from other locations that Indigenous stories are appealing to tourists (Wright *et al.*, 2009; Walker and Moscardo, 2016). While this seems like a positive opportunity to make tourism on MI more inclusive, one must remember that Australian Indigenous communities have complex rules about the ownership of, and rights to tell and hear stories, and the use of Indigenous stories requires careful consultation (Sentina *et al.*, 2017). Not all Indigenous communities want to share their traditional stories, and it is important that Indigenous stories not be used in tourism as a way to avoid deeper and more difficult challenges in engaging in more inclusive tourism (Hugger, 2020). Beyond Indigenous members of the MI community, there are also others who may, or may not, want to share their stories with tourists, and care needs to be taken when deciding whose stories to tell and whose stories to keep private. Censorship of stories could become a problematic issue in diverse communities.

Storytelling has had a central role in tourist experiences because stories are both enjoyable and persuasive. Stories combine all three of Aristotle's core elements required for persuasion: logic, credibility and emotional connection (Dolan and Naidu, 2013). Stories work as a tool

Table 11.2. Examples of the main types of storytelling experiences for tourists.

Storytelling method	General examples	MI examples
Tour guides/ professional storytellers	Croatian storytelling initiative, Istra In Spirit, which works with destination residents to offer story-themed tours and drama productions based on local stories (https://www.istrainspirit.hr/en/)	Potential to develop a programme to train local residents as tour guides telling MI stories
Audio/visual presentations/ interpretive text/ exhibits in built attractions	The story of chocolate, presented at the Chocolate Museum in Cologne, Germany (https://www.schokoladenmuseum.de/en/)	Potential to create various audio/ visual story presentations housed at key transport nodes on MI
Self-guided trails (working, driving and snorkelling)	Self-guided driving Ship Wreck Trail on Kangaroo Island in Australia (https://www.environment.sa.gov.au/ topics/heritage/maritime-heritage/visiting-maritime-heritage-places/shipwreck-trails/ Kangaroo_Island)	Already a World War II story told through interpretive signage on the Forts Walk (https://www.queensland.com/in/ en/plan-your-holiday/journeys/p-56b262eb7b935fbe730e2931-forts-walk-magnetic-island-national-park) Potential to tell other stories in a similar fashion on other walking trails and on the existing snorkelling trail with stories in these locations focusing on themes related to the two PAs linked to MI An underwater art installation as part of the Museum of Underwater Art has been proposed for MI with the artist involved typically using these sculptures to tell stories related to environmental conservation and sustainability issues (https://www.moua.com.au/visit/ magnetic-island/)
Mobile apps/ websites/virtual reality experiences	Use of QR codes on signs placed along a discovery trail to link Chinese tourists to story-based podcasts in Chinese. Green Island National Park in Australia (https:// www.tropicnow.com.au/2019/july/31/ island-technology-used-to-overcome-cairns-tourism-language-barriers)	Potential to use online and mobile technologies to tell any stories from the storyworld described in Table 11.1

for creating rewarding and educational experiences for three main reasons:

- stories are a basic cognitive structure, which makes them very easy to follow and understand (Shrum and Lee, 2012; Green and Dill, 2013);
- stories are seen as entertainment rather than persuasion and so they do not trigger counter-arguing responses, which allows people to focus their attention on any messages that might be in the story (Carpenter and Green, 2012; Nabi and Meyer-Guse, 2013); and
- the emotional elements in stories allow for personal connections and encourage active engagement (Nabi and Green, 2015).

Not surprisingly, stories are often used in heritage interpretation in natural PAs to develop a conservation ethic and support responsible behaviour among visitors to these environments. Research into the persuasive power of stories also indicates that personal relevance, personal connections and active engagement can further enhance their ability to create positive tourism experiences (Lebowitz and Klug, 2011; Moscardo, 2017).

Creating your own personal story is an example of adding these personal connections and active engagement. There is evidence that tourists do seek to engage in certain types of stories while they travel (Woodside, 2010; Woodside and Martin, 2015), especially stories of adventure, personal challenge, self-discovery and nostalgia (Hennig, 2002). At the most basic level, the tourist is actively and overtly framing their experience as an adventure and requires little more than access to the setting. Hawkins's (2013) study of Australian tourists trekking the Kokoda Trail in Papua New Guinea notes the prevalence of tourists describing the trip as an opportunity to challenge themselves and to live out a story of personal adventure.

Tourism providers can, however, facilitate personal story-based experiences by providing more than just access to a setting. They can also provide props and other characters, as well as direction and production support for the tourists' story creation (Mossberg, 2008). Tourists could, for example, be encouraged to recreate journeys taken by historical and fictional characters using annotated maps or guidebooks. Mobile urban dramas provide an example of more complex personal story co-creation opportunities. Hansen *et al.* (2012) describe several of these dramas, which combine story creation with mobile technologies to allow tourists to have a destination experience built around a fictional story. *Hikuin's Vendetta*, for example, was created to support a self-guided Viking history themed tour of the destination based on a murder mystery set in the year 1049. The tourist takes on the role of Hikuin, who is trying to solve the murder of his father. The tourist moves around to different locations gathering information and making decisions that guide the plot development. Each tourist thus creates their own personal version of the story, while simultaneously moving through the destination and learning about the historical features of the destination. Another example described by Hansen *et al.* (2012) is 'Hasleinteractive', a conservation themed story and treasure hunt style experience aimed more at children. In this example, the story is a science fiction thriller, and the young tourists take on the role of assisting two fictional scientists to find out why large areas of nature are dying. In this experience, the tourists move through the natural environment seeking evidence and conducting tests. Local residents can engage with tourists in these story-based tourist experiences in several ways. Firstly, they can be storytellers (see the Croatian example in Table 11.2) as either guides or as people called upon to tell their personal stories as part of other structured tourist activities. Tours of Soweto, in South Africa, often include locals, both as the overall tour guide who tells their own personal story of Soweto as part of the tour, and as people that are met throughout the tour and given opportunities to tell their personal stories at key locations and stops, such as lunch. Secondly, storytelling experiences can be developed around the personal stories of residents. The Murmur project, available in several cities including Toronto and Dublin (DeletedUser, 2011), provides an example of making local resident stories available for tourists through a mobile app.

Harrington (2004) interviewed both MI residents and regular visitors from the nearby regional centre of Townsville about their identity and the role of MI in that identity. She concluded that the two PAs linked to MI were central to both their image of MI and why they, as individuals, felt such a strong connection to MI. The importance of the PAs to MI resident perceptions of their quality of life was also a dominant theme in the tourism futures scenario exercises discussed earlier. It is likely that local stories told to tourists will be linked to these PAs and their importance to both residents and the world beyond MI. Several existing tour operators are local residents seeking to share their personal story of MI and provide tourist access to the PAs that are so central to this location (see https://www.aquascenecharters.com.au/the-aquascene-story.html). Such opportunities could be expanded with training and support from PA managers. The Brecon Beacons National Park Ambassador Scheme, in Wales, UK, provides an example of how such a programme could be structured (https://www.greentraveller.co.uk/post/brecon-beacons-ambassadors).

A common theme in these discussions of storytelling about MI is the importance of the adjacent PNAs, environmental conservation and sustainability to the personal stories of the people who live in this gateway community. In some of these stories, it is the efforts of the individuals involved in engaging in conservation actions that could feature in MI conservation and sustainability stories. The work of two marine scientists, Rick Braley and Charlie Vernon, often features examples of direct conservation action taken on or around MI. In others, sustainability emerges as an outcome of some other activity. The Netflix documentary series *Izzy's Koala World* follows Izzy Bee and her mother, a veterinarian on MI, as they rescue and rehabilitate injured wildlife, especially koalas. The presentation of these animal rescue and bonding stories typically includes discussions of how the animals were injured or became endangered, and how humans can change their behaviour in order to help with animal conservation in the area. Finally, stories of everyday life and adventures as a resident can include both elements of how MI residents make the most of the natural environment around them, and how they have adapted their daily lives to support greater sustainability and conservation.

These personal stories of MI can also be used to identify themes, or plots, for story-creating experiences. Harrington's (2004) work suggested that local resident stories of MI were often about:

- personal challenges and adventure linked to the importance of learning to live with the lack of urban facilities, such as paved roads and clearly marked trails;
- the sense of escape to simpler, and in recent years, more sustainable lifestyles; and
- the magical qualities of this particular island.

Such themes could be used to reframe tourist activities as the creation of personal stories. For example, presenting the MI walking trails, many of which are in the national park, as a potential story of personal challenge, discovery or adventure.

Finally, local residents could be involved in story-creating experiences as service providers (equipment, transport and food), producers (offering maps and suggested itineraries) or other characters in more formally presented gamified story-based experience opportunities. The key here is to shift the promotional rhetoric for MI from 'relaxing on a scenic, sunny island to escape normal life' to 'actively engaging in adventures and learning, especially about sustainability, to create stories worth telling others'.

This reframing around a local story-based approach to tourist experience design is not just an opportunity to engage locals more directly in a positive way in tourism but is also likely to contribute to the long-term resilience and sustainability of tourism to this gateway community. Moreover, story creation is likely to increase repeat visitation and offer more appeal to tourists coming from the local region. Repeat business from local markets is a key element in reducing carbon emissions from travel, and these local markets are more likely to be able to visit when the larger tourism distribution system is impacted by crises. As has been previously noted, this was evident in 2020 during the first year of the Covid-19 pandemic. Except for a five- to six-week break, when all residents in the state were under stay at home orders, local tourists provided considerable support for businesses on MI, even when state and national borders remained closed.

11.4 Final Considerations

Some of the challenges for tourism businesses seeking to attract repeat local business is to find ways to offer variety in experiences and to encourage local tourists to see destinations as offering something they do not already know. Telling stories about hidden worlds in local places and finding the exotic underneath the mundane is one way to achieve this. Richards (2020) has argued that we need to offer tourists the opportunity to be more creative, and story creation is a potential pathway to this outcome.

It is not just tourists, however, that should be encouraged to be creative. This chapter has argued that tourism planners and developers need to be more creative and find new stories of tourism itself, in collaboration with local residents. In the case of PAs, managers also need to be more imaginative and flexible in the way they think about the stories they present as a means to justify the existence and conservation of PAs. PA managers have often presented their places within a western scientific framework, arguing for the importance of various natural, and sometimes cultural, heritage values. The overwhelming evidence from sociology, anthropology and psychology is that humans are storytelling animals (Gottschall, 2012), and that values are best transmitted through personal, emotional, archetypal stories. This challenges these managers to become better storytellers and to find new connections between residents, their stories, and the places that they live in and/or near.

References

Adaval, R. and Wyer Jr, R. (1998) The role of narratives in consumer information processing. *Journal of Consumer Psychology* 7(3), 207–245.

Amer, M., Daim, T.U. and Jetter, A. (2013) A review of scenario planning. *Futures* 46, 23–40.

Amy, S. (2016) The art of participation: the case of creative communities in Indonesia. *Community Development Journal* 52(1), 171–185.

Burnam-Fink, M. (2015) Creating narrative scenarios. *Futures* 70, 48–55.

Carpenter, J.M. and Green, M.C. (2012) Flying with Icarus. In: Shrum, L.J. (ed.) *The Psychology of Entertainment Media*, 2nd edn. Routledge, New York, NY, USA, pp. 169–194.

Cowey, P. (2017) Performing planning: understanding community participation in planning through theatre. *The Planning Review* 88(4), 401–421.

Da Costa, D. (2013) *Development dramas*. Routledge, New Delhi, India.

De Massis, A. and Kotlar, J. (2014) The case study method in family business research: guidelines for qualitative scholarship. *Journal of Family Business Strategy* 5(1), 15–29.

DeletedUser (2011) *Hello City? It's me, Lisa*. Available at: https://challenges.openideo.com/challenge/vibrant-cities/inspiration/-hello-city-it-s-me-lisa-#comments-section

Dolan, G. and Naidu, Y. (2013) *Hooked*. Wiley, Melbourne, Australia.

Dunbar, R.I.M., Teasdale, B., Thompson, J., Budelmann, F., Duncan, S. *et al.* (2016) Emotional arousal when watching drama increases pain threshold and social bonding. *Royal Society Open Science* 3(9), 1–11. DOI: 10.1098/rsos.160288.

Dunn, C.D. (2017) Personal narratives and self-transformation in postindustrial societies. *Annual Review of Anthropology* 46, 65–80.

Fivush, R. (2011) The development of autobiographical memory. *Annual Review of Psychology* 62, 559–582. DOI: 10.1146/annurev.psych.121208.131702.

Goldstein, E.B. (2015) *Cognitive Psychology: Connecting Mind, Research, and Everyday Experience*, 4th edn. Cengage, Stanford, CT, USA.

Gottschall, J. (2012) *The Storytelling Animal*. Houghton Mifflin Harcourt, New York, NY, USA.

Green, M.C. and Dill, K.E. (2013) Engaging with stories and characters. In: Dill, K.E. (ed.) *The Oxford Handbook of Media Psychology*. Oxford University Press, Oxford, UK, pp. 449–461.

Grimwood, B.S., Stinson, M.J. and King, L.J. (2019) A decolonizing settler story. *Annals of Tourism Research* 79, 102763.

Hall, C.M. (2008) *Tourism Planning*, 2nd edn. Pearson, Harlow, UK.

Hansen, F.A., Kortbek, K. and Grønbæk, K. (2012) Mobile urban drama: interactive storytelling in real world environments. *New Review of Hypermedia and Multimedia* 18(1-2), 63–89.

Harrington, J.T. (2004) *'Being here': heritage, belonging and place making: a study of community and identity formation at Avebury (England), Magnetic Island (Australia) and Ayutthaya (Thailand)*. Doctoral dissertation. James Cook University, Townsville, Australia.

Hawkins, J. (2013) What better excuse for a real adventure. *Public History Review* 20, 1–23.

Hennig, C. (2002) Tourism: Enacting modern myths. In: Dann, G.M.S. (ed.) *The Tourist as a Metaphor of the Social World*. CAB International, Wallingford, UK, pp. 169–188.

Hewlett, D. and Edwards, J. (2013) Beyond prescription: community engagement in the planning and management of national parks as tourist destinations. *Tourism Planning & Development* 10(1), 45–63.

Higgins-Desbiolles, F. (2020) The "war over tourism": Challenges to sustainable tourism in the tourism academy after COVID-19. *Journal of Sustainable Tourism* 29(4), 551–569. DOI: 10.1080/09669582.2020.1803334.
Hugger, D. (2020) *Avoiding the Trap of Cultural Tokenism*. Available at: https://www.cela.org.au/2020/06/16/cultural-tokenism/ (accessed 7 February 2022).
Kent, M. (2015) The power of storytelling in public relations. *Public Relations Review* 41(4), 480–489.
Koens, K., Postma, A. and Papp, B. (2018) Is overtourism overused? Understanding the impact of tourism in a city context. *Sustainability* 10(12), 4384.
Laszlo, J. (2008) *The Science of Stories*. Routledge, London, UK.
Lebowitz, J. and Klug, C. (2011) *Interactive Storytelling for Video Games*. Elsevier, Amsterdam, The Netherlands.
Moscardo, G. (2011) The role of knowledge in good governance for tourism. In: Laws, E., Richins, E.H., Agrusa, J., and Scott, N. (eds) *Tourist Destination Governance*. CAB International, Wallingford, UK, pp. 67–80.
Moscardo, G. (2017) Stories as a tourist experience design tool. In: Fesenmaier, D.R. and Xiang, Z. (eds) *Design Science in Tourism*. Springer International, Basel, Switzerland, pp. 97–124.
Moscardo, G. (2018) Tourist experience design: a storytelling framework. In: Cai, L.P. and Alaedini, P. (eds) *Quality Services and Experiences in Hospitality and Tourism*. Emerald, Bingley, UK, pp. 93–108.
Moscardo, G. (2019a) Rethinking the role and practice of destination community involvement in tourism planning. In: Andriotis, K., Stylidis, D. and Weidenfeld, A. (eds) *Tourism Policy and Planning Implementation*. Routledge, London, UK, pp. 36–52.
Moscardo, G. (2019b) Connecting people with experiences. In: McCool, S. and Bosak, K. (eds) *A Research Agenda for Sustainable Tourism*. Edward Elgar, Cheltenham, UK, pp. 70–89.
Moscardo, G. (2020a) Stories and design in tourism. *Annals of Tourism Research* 83. DOI: 10.1016/j.annals.2020.102950.
Moscardo, G. (2020b) The story turn in tourism: forces and futures. *Journal of Tourism Futures* 7(2), 168–173. DOI: 10.1108/JTF-11-2019-0131.
Moscardo, G. and Murphy, L. (2013) There's no such thing as sustainable tourism: Reconceptualizing tourism as a tool for sustainability. *Sustainability* 6(1), 2538–2561.
Moscardo, G. and Murphy, L. (2015) Educating destination communities for sustainability in tourism. In: Moscardo, G. and Benckendorff, P. (eds) *Education for Sustainability in Tourism*. Springer, New York, NY, USA, pp. 135–154.
Moscardo, G., Blackman, A. and Murphy, L. (2014) Tourism development as Greek tragedy. In: *Politics, Policy and Governance in Sustainable Tourism*. BEST EN Think Tank XIV, 23-26 June 2014, Ljubljana, Slovenia, pp. 217–231.
Mossberg, L. (2008) Extraordinary experiences through storytelling. *Scandinavian Journal of Hospitality and Tourism* 8(3), 195–210.
Moyle, B.B., Croy, G. and Weiler, B. (2010) Community perceptions of tourism: Bruny and magnetic Islands, Australia. *Asia Pacific Journal of Tourism Research* 15(3), 353–366.
Nabi, R.L. and Green, M.C. (2015) The role of a narrative's emotional flow in promoting persuasive outcomes. *Media Psychology* 18(2), 137–162.
Nabi, R.L. and Meyer-Guse, E. (2013) The psychology underlying media-based persuasion. In: Dill, K.E. (ed.) *The Oxford Handbook of Media Psychology*. Oxford University Press, Oxford, UK, pp. 285–301.
Nunkoo, R. (2015) Tourism development and trust in local government. *Tourism Management* 46, 623–634.
Oteros-Rozas, E., Martín-López, B., Daw, T.M., Bohensky, E.L., Butler, J.R. *et al.* (2015) Participatory scenario planning in place-based social-ecological research. *Ecology and Society* 20(4), 32.
Polletta, F., Chen, P., Gardner, B. and Motes, A. (2011) The sociology of storytelling. *Annual Review of Sociology* 37, 109–130.
Popova, Y.B. (2015) *Stories, Meaning, and Experience*. Routledge, Abingdon, UK.
Queensland Parks & Forest (2020) *About Magnetic Island National Park*. Available at: https://parks.des.qld.gov.au/parks/magnetic-island/about (accessed 7 February 2022).
Queensland Times (2020) *Magnetic Island Proves a Treat in COVID-19 Recovery*. Available at: www.qt.com.au
Richards, G. (2020) Designing creative places: the role of creative tourism. *Annals of Tourism Research* 38(4), 1225–1253.
Ruhanen, L. (2013) Local government: facilitator or inhibitor of sustainable tourism development? *Journal of Sustainable Tourism* 21(1), 80–98.

Saarinen, J., Rogerson, C.M. and Hall, C.M. (2017) Geographies of tourism development and planning. *Tourism Geographies* 19(3), 307–317.

Sentina, M., Mason, E. and Janke, T. (2017) *Legal Protection of Indigenous Knowledge in Australia*. Australian Government IP, Canberra, Australia.

Shrum, L.J. and Lee, J. (2012) The stories TV tells. In: Shrum, L.J. (ed.) *The Psychology of Entertainment Media*, 2nd edn. Routledge, New York. NY, USA, pp. 147–168.

Snyman, S. and Bricker, K.S. (2019) Living on the edge. *Journal of Sustainable Tourism* 27(6), 705–719.

Stevenson, N. (2019) Developing cultural understanding through story-telling. *Journal of Teaching in Travel & Tourism* 19(1), 8–21.

Stoker, P., Rumore, D., Romaniello, L. and Levine, Z. (2020) Planning and development challenges in western gateway communities. *Journal of the American Planning Association* 87(1), 21–22.

Tourism Queensland (2009) *Townsville North Queensland Tourism Opportunity Plan 2009-2019*. Tourism Queensland, Brisbane, Australia.

Volo, S. (2010) Bloggers' reported tourist experiences. *Journal of Vacation Marketing* 16(4), 297–311.

Walker, K. and Moscardo, G. (2016) Moving beyond sense of place to care of place: the role of Indigenous values and interpretation in promoting transformative change in tourists' place images and personal values. *Journal of Sustainable Tourism* 24(8-9), 1243–1261.

Woodside, A.G. (2010) Brand-consumer storytelling theory and research. *Psychology and Marketing* 27(6), 531–540.

Woodside, A.G. and Martin, D. (2015) Introduction: the tourist gaze 4.0. *International Journal of Tourism Anthropology* 4(1), 1–12.

Wright, S., Suchet-Pearson, S., Lloyd, K., Burarrwanga, L.L. and Burarrwanga, D. (2009) 'That means the fish are fat': Sharing experiences of animals through Indigenous-owned tourism. *Current Issues in Tourism* 12(5-6), 505–527.

Yeoman, I. and Postma, A. (2014) Developing an ontological framework for tourism futures. *Tourism Recreation Research* 39(3), 299–304.

Yin, R.K. (2012) *Applications of Case Study Research*, 3rd edn. Sage, Los Angeles, CA, USA.

12 Sustainable Tourism Initiatives and Impediments: The National Mall as an Urban Gateway Community

Margaret J. Daniels* **and Susan L. Slocum**
George Mason University, Virginia, USA

Abstract
Sustainable resource management within urban gateway communities can be fraught with conflict between local and national governance structures, site managers and users, in particular when funding and human capital are insufficient to address the complex needs of protected areas (PAs). The purpose of this chapter is to thematically review sustainability and PA management plans and practices within the National Mall and Memorial Parks in light of the surrounding gateway community of Washington, District of Columbia (DC). Ten Green Parks Plan goals, developed by the National Park Service (NPS), provide a basis of examination, with grounded theory and qualitative document analysis used to deduce specific challenges and notable enhancements. Data sources include printed information, consultancy, formal interviews, informal conversations and/or presentations by key stakeholders. Since the needs of residents and tourists are inevitably linked, this analysis serves as a model for contextualizing national sustainability guidelines on a local level for urban gateway destinations.

Keywords: urban gateway communities, National Park Service (NPS), National Mall and Memorial Parks, sustainability

12.1 Introduction

Urban tourism is often overlooked in literature specific to protected area (PA) management. Yet, the cultural and natural assets that drive tourism visitation in urban cores often fall within the realm of PAs, and these features face many of the same governance challenges as rural PAs. Overcrowding, resource degradation and deferred maintenance are common to urban destinations, impacting residents and visitors in ways that, at times, intersect, yet often diverge, and lead to user conflicts.

Iconic features may be a primary draw for urban visitors, such as the Acropolis of Athens, the Great Pyramid of Giza and the Taj Mahal in Agra. Each of these destinations is surrounded by a heterogeneous gateway community, where gateway communities have been defined as areas that border and serve as entry points to a nation's public lands and parks (Frauman and Banks, 2011). These resources have multifaceted cultural significance that may imbue different meanings to local residents, national citizens and international visitors based on their interpretations of the associated destination's history.

*Corresponding author: mdaniels@gmu.edu

 Tourism Transformations in Protected Area Gateway Communities (eds S.L. Slocum *et al.*)
DOI: 10.1079/9781789249033.0012

When an urban centre is also a capital city, resource management can be fraught with tension, as new projects tend to garner more attention than the maintenance of long-standing features and infrastructure. 'The juxtaposition of symbolic meaning and resource degradation poses a struggle to which managers of any heritage site facing funding shortages can most likely relate' (Daniels *et al.*, 2015, p. 2). In the United States, Washington, District of Columbia (DC) is one such example. While estimates vary, DC hosts roughly 32 million visitors from around the world per year, the majority of which are leisure travellers (NPS, 2020a). With a resident population of just over 705,000 (DC. Gov, 2020), the residents of this capital city are far outnumbered by guests, yet the locals must contend with the day-to-day impacts of neglect in areas such as greenspace and transportation.

A primary motivation for visiting DC is to spend time within the National Mall and Memorial Parks (National Mall, Fig. 12.1). In addition to iconic monuments and memorials, such as the Washington Monument, World War II Memorial, Vietnam Veterans Memorial and Martin Luther King, Jr. Memorial, visitors are drawn to the National Mall greenspace, the lawns of which stretch east to the US Capitol and west to the Lincoln Memorial. This historic landscape and the enclosed Lincoln Memorial Reflecting Pool act as working cultural symbols of the American ideals of democracy, opportunity and First Amendment freedom of speech demonstrations. Hallmark events, such as a National Cherry Blossom Festival, the National Independence Day Parade and the US Presidential Inauguration, take place within and around the confines of the National Mall, attracting millions of visitors.

The National Mall, as a whole, encompasses approximately 1000 acres that are managed by the National Park Service (NPS), a bureau within the US Department of the Interior (NPS, 2017a). The origins of the National Mall date back to the vision and plans of Pierre L'Enfant, who was commissioned by President George Washington in 1791 (NPS, 2017a). With over 80 structures and 150 parkland areas, the historic spaces are both awe-inspiring and functional (Daniels *et al.*, 2015; NPS, 2017a), leaving them subject to significant wear and tear.

For over 30 years, NPS has funded research designed to measure levels of acceptable change brought about by tourism visitation and recreational use within its park boundaries (e.g., Stankey *et al.*, 1985). By the late 1990s,

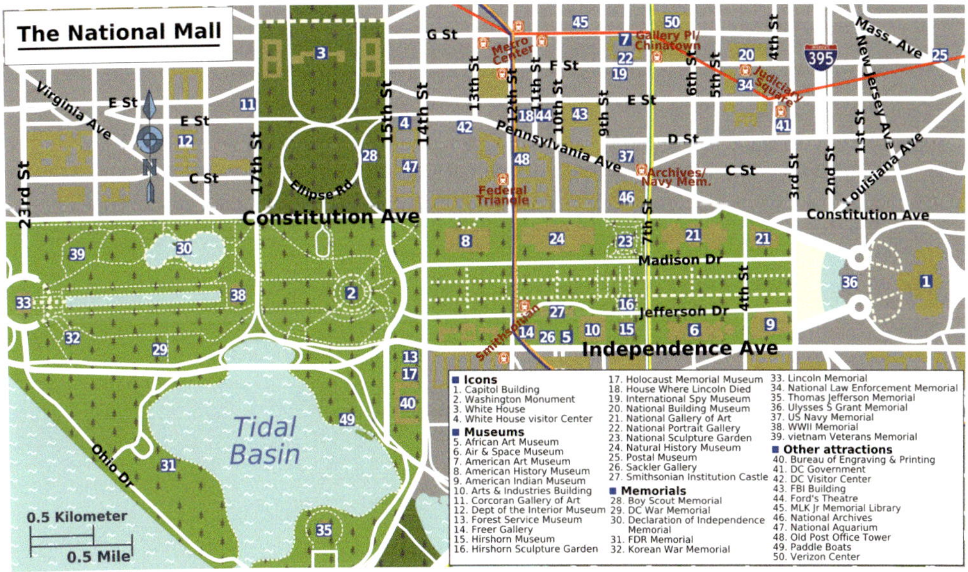

Fig. 12.1. National Mall and Memorial Park map (Peter Fitzgerald – this file is licensed under the Creative Commons Attribution-Share Alike 3.0, unported licence).

a majority of superintendents of NPS holdings across the US were reporting harmful effects on water quality, air quality and vegetation within park boundaries caused by visitation, including within the National Mall (see Wang and Miko, 1997). These impacts affect locals who reside in close proximity to parks, in particular in gateway communities with dense resident populations (Frauman and Banks, 2011).

In response to widespread calls for sustainable management, NPS released a comprehensive Green Parks Plan in 2012, which was then updated in 2016 in line with its 100th anniversary (NPS, 2016). The revised Green Parks Plan, 'reaffirms the NPS vision to integrate sustainable practices into every aspect of Service [*sic*] operations and outlines updated objectives to achieve that long-term vision' (NPS, 2016, p. 3). The ten strategic goals included in the plan are specific to meeting environmental performance standards, reducing greenhouse gas emissions, improving energy performance, improving water use efficiency, adopting green transportation, increasing recycling standards, promoting outdoor experiences, strengthening sustainability partnerships, fostering visitor engagement in sustainable practices and enhancing sustainable landscapes (Table 12.1). Slocum (2017) criticized the scope of the plan, arguing that 'the focus on internal operations is counter to sustainability principles where the economy, environment, and society are dynamic and interrelated elements that require research-informed governance' (p. 1854). An investigation of these macro-level goals within the unique confines of a given NPS unit can shed light on their utility and differential application.

When considering the Green Parks Plan goals within the context of the National Mall, these ideals must be balanced with the US

Table 12.1. The Green Parks Plan goals (NPS, 2016, pp. 7–16).

Goal	Description	Application to National Mall sustainability themes
1. Continuously improve environmental performance	The NPS will meet and exceed the requirements of all applicable environmental laws and adopt sustainable best practices in all facility operations	Historic landscapes Water Transportation
2. Be climate friendly and climate ready	The NPS will reduce GHG emissions and adapt facilities at risk from climate change	Transportation
3. Be energy smart	The NPS will improve facility energy performance and increase reliance on renewable energy	Transportation
4. Be water wise	The NPS will improve facility water use efficiency	Water
5. Green our rides	The NPS will transform our fleet and adopt greener transportation methods	Transportation
6. Buy green and reduce, reuse and recycle	The NPS will purchase environmentally friendly products and increase waste diversion and recycling	Water
7. Preserve outdoor experiences, promote healthy engagement	The NPS will promote healthy outdoor experiences and minimize the impact of facility operations on the environment	Historic landscapes Transportation
8. Strengthen sustainability partnerships	The NPS will incorporate sustainability initiatives into new and existing partnerships	Water Transportation
9. Foster sustainability beyond our boundaries	The NPS will engage visitors about sustainability and invite their participation	Historic landscapes Water Transportation
10. Green our grounds	The NPS will enhance the sustainability of our landscapes	Historic landscapes

Constitution. An important provision to the First Amendment is the fact that the right for freedom of expression guarantees access to public gathering spaces. For example, provisions taken to enhance sustainable landscapes and promote healthy soils must not restrict visitors from utilizing the National Mall greenspace. Adhering to sustainable landscape maintenance guidelines, when public gatherings must take precedence, poses particularly complex challenges to land stewardship. As such, the purpose of this chapter is to thematically review sustainability and PA management plans and practices within the National Mall in light of the surrounding gateway community of DC. The ten Green Parks Plan goals will provide a basis of examination, with grounded theory and qualitative document analysis used to deduce specific challenges and notable enhancements.

12.2 Literature Review

Tourism in urban core destinations supports economic functions, such as job creation, infrastructure development and the provision of services, that enrich the quality of life for residents. Urban tourism is categorized as having 'concentrated resources, high culture influences, rich activity contents, strong inclusiveness, multiple function elements, distinct geographical advantages, and convenient transportation' (Zheng-Xin and Pei, 2014, p. 463). Cities are subject to high tourism volumes, acting as destinations in their own right and as entryways to neighbouring attractions. When combined with dense residential populations, urban tourism results in extensive degradation to the natural environment that can be expensive and complicated to mitigate because of conflicting governance structures, heterogeneous community groups and tourism demand that frequently overshadows local needs (Muangasame and McKercher, 2015).

While the issue of sustainability is salient to urban policy-makers, a specific focus on tourism is often neglected, primarily because tourism encompasses a multitude of diverse industry sectors (Moyle *et al.*, 2014). Sustainability requires both political will and community engagement, and participatory constraints can be exacerbated in urban areas based on income disparities, ethnic diversity and transient populations (Muangasame and McKercher, 2015). Moreover, a destination's brand and focal areas are often conferred to destination marketing organizations (DMOs) whose primary function is to increase tourism (Pearce, 2015). DMOs are often at the forefront of lobbying activities that focus on growth and may eschew restrictive policies that address sustainability issues or encourage visitor management plans (Kavaratzis, 2017).

Sustainable development in urban areas requires extensive financial investments, and residents may object to tax revenue being spent on tourism services or tax incentives to tourism businesses that implement sustainability strategies (Kerr, 2003). At the same time, residents of, and visitors to, urban destinations are highly cognizant of the symbolism inherent to capital cities, and cognitive dissonance occurs when the spaces therein are in a state of disrepair (Harmon *et al.*, 2011; Daniels *et al.*, 2015). While key attractions may dictate and geographically delimit supply and demand to an extent (see Fava and Rubio, 2017), the degree to which supporting sectors (e.g. food/beverage, lodging, retail) are readily accessible may lead to an expansion of geographical demarcations.

What constitutes sustainability in urban destinations may differ significantly from the tenets applied in rural or remote natural areas. Sustainable tourism initiatives are seen to be most closely aligned with nature-based destinations, while the culturally-oriented focus of urban tourism does not readily evoke the associated imagery; as such, policy-makers may struggle to define the role of sustainability in urban tourism planning (Maxim, 2015). For example, Maxim (2016) found that sustainable tourism practices in London boroughs might include the promotion of public transport, facilitation of temporary art activities in vacant properties, encouragement of local hiring practices and implementation of measures to reduce waste and water consumption. Similarly, Bellini *et al.* (2017) recognize infrastructure, specifically public transport and traffic management, as a key avenue to support sustainable urban tourism. As these examples suggest, some sustainability issues may be unique to, or exacerbated within, urban destinations.

Childers *et al.* (2014) describe two categories of sustainability solutions: those that 'tweak' current systems versus those that transform current systems by finding new ways of doing business. The development of gateway destinations incorporates both approaches. Urban national parks in capital cities, for instance, generally have more immediate access to stakeholder resources when compared to remote destinations. In DC, resources available to the National Mall extend well beyond the NPS, and may include cooperative agreements, consultation and financial support provided by entities, such as the National Capital Planning Commission, the Trust for the National Mall, the Advisory Council on Historic Preservation, the Downtown DC Business Improvement District, the US Department of Transportation, the US Environmental Protection Agency, Destination DC, the Smithsonian Institution and the Metropolitan Washington Council of Governments, just to name a few (Daniels *et al.*, 2015). As noted by Fuller (2016), securing investment may be more successful when federal and local agencies work together and pool resources. At the same time, addressing competing needs becomes exponentially more complicated as the list of powerful stakeholders grows, and responsibility in relation to tourism impacts and mitigation may shift based on focal areas of interest (Brash *et al.*, 2011).

As a whole, the 85 million acres of the NPS are underfunded, with US$13.1 billion in deferred maintenance as of 2019 in key areas, such as road maintenance, buildings and waterfront systems (Pew, 2020). The National Mall has the highest dollar amount of deferred maintenance of all NPS holdings, at $840 million, with overdue repairs and refurbishment needed for crumbling infrastructure and decimated greenspace, as well as a dearth of basic amenities, such as restrooms, seating and water fountains (Pew, 2020). From an operations standpoint, the National Mall requires more electricity and potable water than any other NPS holding (NPS, 2010b).

Faced with the reality of significant budget shortfalls, the NPS has still garnered extensive research related to sustainability planning and implementation. Funding for such research and associated initiatives requires proof of competitive advantage, cost savings, societal relevancy and human resource support (Eagles, 2014; Slocum, 2017). Congressional gridlock and restrictive hiring policies have resulted in a human resources crisis for the NPS, with position backlogs of up to 40% per unit – 'acting' superintendent positions that result in a revolving door of leaders who do not have sufficient time to become invested in individual park projects before moving on, hiring investigations that require a year or more to fill a single position and chronic overreliance on volunteers to complete critical park functions (Repanshek, 2019). As such, external partnerships are central to the completion of sustainability research.

12.3 Methodology

The fundamentals of grounded theory and qualitative document analysis are applied to the investigation of extant National Mall research projects and supporting materials that focus, at least in part, on issues of sustainability and PA management. Grounded theory allows researchers to gain insight through systematic data review and analysis of conceptual documentation (Hardy, 2005). Theory is developed from printed information, consultancy, formal interviews, informal conversations and/or presentations by key stakeholders. The goal is to move from inductively driven data to deductive patterns and trends (Daymon and Holloway, 2002). Concurrently, qualitative document analysis 'entails finding, selecting, appraising (making sense of), and synthesising data' (Brown, 2009, p. 28) from printed and electronic material to elicit themes, categories and examples from the content, while seeking convergence and corroboration from different data sources (Brown, 2009).

The data-based material reviewed included: primary research studies that supported the two-volume National Mall Plan (NPS, 2010b), as well as the associated environmental impact statement, best management practices at urban parks report, and implementation priorities for the National Mall; National Mall records of determination (RODs) specific to sustainability issues that provide the rationale for closures or public use restrictions, as relevant at this writing (February, 2021); and National Mall regulations specific to sustainability that are derived from the *Superintendent's Compendium* (NPS, 2018a), which lists additional restrictions and designations that are in place based on the superintendent's discretionary authority, again as in place at the time of this writing. As the National Mall Plan includes ambitious long-term goals, the sustainability specifics outlined in this document, and the associated supporting materials, are used as a baseline, with the RODs and *Superintendent's Compendium* reviewed as examples of status updates to key priorities within the National Mall Plan that predate the RODs and *Superintendent's Compendium*. In order to delimit the findings to primary-level initiatives that allow for park-to-park comparisons, multi-purpose large scale infrastructure plans were not considered in the current analysis. A critical lens was applied to extract sustainability issues as related to the National Mall and the surrounding gateway DC community by reviewing the extent to which the resulting recommendations converge or diverge with the NPS Green Parks Plan goals.

12.4 Results

Three dominant sustainability themes emerged from the analysis of the data sources. The themes encompassed PA guidelines, usage policies and implementation priorities specific to: historic landscapes (turf, soil, trees, plants); water (seawalls, shorelines, water-based recreation, potable water); and ground transportation and circulation (tour buses, site seeing transport, pedestrian/bicycle circulation). Embedded in these themes are the ten Green Parks Plan goals, which will be identified as relevant to each theme, as well as gateway destination considerations.

12.4.1 Historic landscapes

Sustainable greenspace management is essential on the National Mall to address the impacts of compaction and erosion while understanding that the landscapes will need to serve high visitor use. Urban parklands, such as Central Park (New York City), Golden Gate State Park (San Francisco) and The Royal Parks (London), were analysed as points of comparison in setting forth best management practices for the National Mall (NPS, 2007). The resulting landscape standards for the National Mall further acknowledged that parklands in capital cities must balance landscape rehabilitation and resource conservation with constitutionally protected access and national-level events that, while resulting in significant landscape deterioration, are essential to the promotion and 'understanding of national culture, democratic principles, diversity, or identity' (NPS, 2007, p. 9). Landscape rehabilitation projects are listed within the high priority National Mall Plan implementation guidelines and echoed in the Green Parks Plan goals to include 'Continuously improve environmental performance', 'Green our grounds', 'Preserve outdoor experiences, promote healthy engagement' and 'Foster sustainability beyond our boundaries'.

As part of visitor outreach, National Mall ranger programmes, wayside exhibits and interpretive materials are designed to educate visitors regarding the impacts of their actions on protected natural resources, with varying levels of success. For instance, the annual National Cherry Blossom Festival is the largest and longest-held event within the National Mall calendar and is reliant on the historic cherry trees that line the Tidal Basin trail (Daniels, 2020). While it is illegal to climb the trees, step on the exposed roots, remove the branches or pick the blossoms, and highly visible educational materials are strategically placed to consistently inform visitors regarding these policies, the fact remains that 'visitors delight in deliberately shaking the trees to create a rainfall of blossoms, with over-zealous adults seeking picture-perfect moments more likely to violate greenspace policies than children' (Daniels, 2020, p. 128). Protection of the cherry trees and other natural resources, while consistently reinforced in the National Mall Plan, requires a level of buy-in from visitors that underpins the challenges facing sustainability-seeking gateway communities. Self-restraint becomes imperative, as enforcement is nearly impossible when visitors outnumber rangers by a ratio of up to 9000 to 1 at popular sites, such as the Lincoln Memorial (NPS, 2010b).

A 2020 ROD addresses turf management within the realm of large-scale gateway-based events, such as presidential inaugurations, with specific emphasis on the essential need to temporarily close delineated turf panels on the National Mall for three-month periods of short-term 'recuperative remediation' as part of a long-term 'multiyear effort to develop a more resilient soil mix' (NPS, 2020b). Importantly, the ROD specifies that the temporary closure would not significantly alter public use patterns and further justifies the necessity of protecting the turf from high impact events to protect the environmental asset (NPS, 2018a). The *Superintendent's Compendium* (NPS, 2018a) also addresses historic landscape preservation directly, with policies specific to limiting access to National Mall turf panels during inclement weather or saturated conditions, the demarcation of recreational activities that are permitted and restricted on the National Mall turf, and public use restrictions specific to areas, such as digging, staking, climbing trees and gathering natural products (e.g., plants, wood, blossoms).

12.4.2 Water

The Tidal Basin, heralded as a 19th-century 'hydrological engineering achievement' (National Trust, 2019, p. 4), based on an innovative design of inlet gates that take in 250 million gallons of water twice a day from the Potomac River during high tides (NPS, 2018b), currently faces such complex sustainability challenges related to daily flooding and crumbling infrastructure that it is now recognized 'as one of America's 11 most endangered historic sites' (p. 15). The adverse conditions have been repeatedly documented yet repairs to the seawalls and walkways have been long delayed due to funding shortages (National Trust, 2020). The Tidal Basin is further impacted by the large ornamental water features on the National Mall, which are reliant on significant volumes of potable water and have algae growth problems associated with poor aeration and circulation (NPS, 2010b). The chemical algaecides utilized affect the regional water quality, as the Tidal Basin's pumping system supports the maintenance of water levels and the features are periodically drained into the Tidal Basin (NPS, 2010b, 2018b). By increasing the amount of non-potable water utilized through methods such as stormwater management, water recirculation and the installation of porous materials that improve water infiltration, potable water consumption can be reduced by 2% annually (NPS, 2010a). Suggested improvements to the Tidal Basin, discussed in additional detail below, reflect the Green Parks Plan goals of 'Continuously improve environmental performance', 'Be water wise', 'Buy green and reduce', 'Reuse and recycle', 'Strengthen sustainability partnerships' and 'Foster sustainability beyond our boundaries'.

The Tidal Basin Ideas Lab was launched in 2019 by the National Trust for Historic Preservation (National Trust) in response to the need for $500 million in repairs and upgrades that were set forth in the National Mall Plan. This collaborative venture includes American Express as a presenting sponsor, the interdisciplinary firm Skidmore, Owings & Merrill as a civic partner and local partnerships with the Trust for the National Mall and NPS. The Tidal Basin Ideas Lab is a call to action that balances 'preservation with the realities of climatic, geological, and social changes' with the goal of preventing the submersion of the Tidal Basin and its monuments, due to flooding caused by climate risk and urbanization (National Trust, para. 2, 2020). Five leading US landscape architecture firms each presented concepts 'to promote an exchange of ideas and to get the public directly involved in the process' (National Trust, para. 3, 2020). Public involvement includes access to the proposals, detailed questionnaires regarding design ideas, a request for visitors' stories and a call for signatures that indicate support of the Tidal Basin pledge.

While there are currently no active RODs specific to the Tidal Basin, the concepts presented and deliberated by the public through the Tidal Basin Ideas Lab are intended to inform future discussions on the upcoming NPS Tidal Basin Master Plan (National Trust, 2020). The *Superintendent's Compendium* details policies and restrictions specific to water-based activities, such as fishing, boating and swimming; for instance, vessel operations along the park shoreline and structures are prohibited except for authorized commercial contracts (NPS, 2018a). The proposals within the Tidal Basin Ideas Lab balance recreational usage with wetland restoration to rebuild the public landscapes and transform the urban ecology (National Trust, 2020).

12.4.3 Ground transportation and circulation

The National Mall Plan addresses the cumulative environmental impacts of visitor transportation, including high priority recommendations designed to 'improve the experience for motorists; equitably treat pedestrians, bicyclists, and drivers; better meet the needs of transit and tour buses; improved convenient transit access along with the visitor transportation system; and enhance the appearance and function of the circulation systems' (NPS, 2010a, p. 481). Improved pedestrian walkways and non-motorized onsite transport, through bicycle rental programmes, are encouraged as green and healthy alternatives to motorized vehicles. The associated preferred alternatives are designed to reduce vehicle emissions, while improving circulation, and reflect the Green Parks Plan goals of 'Continuously improve environmental performance', 'Be climate friendly and climate ready', 'Be energy smart', 'Green our rides', 'Strengthen sustainability partnerships', 'Preserve outdoor experiences, promote healthy engagement' and 'Foster sustainability beyond our boundaries'.

Urban gateway destinations are particularly subject to the environmental impacts of transportation, with road traffic being the most significant contributor to urban air pollution and the heavy-duty diesel engines common to tour buses having 'higher emissions of particulate matter and nitrogen oxides than gasoline engines, leading to adverse effects on human health and the physical environment' as well as contributing to 'traffic congestion, residential neighbourhood disruptions, excessive noise and the obstruction of view corridors and major landmarks' (Daniels *et al.*, 2018, p. 129). With roughly one-third of DC's 32 million annual visitors arriving by tour bus, estimates indicate that the National Mall hosts up to 1200 tour buses a day during the peak spring and summer seasons (Daniels *et al.*, 2018; NPS, 2020). While the primary sites of interest for tourists fall within the boundaries of the National Mall, tour buses regularly infiltrate the larger DC gateway community, as insufficient tour bus parking is available within the confines of the park (Daniels *et al.*, 2018).

Although strict idling laws are in place for commercial vehicles, with fines that start at \$500 and go up to \$4000 for repeat violators of the three-minute idling law (DC Department of Energy & Environment, 2020c), rates of enforcement are minimal based on the sheer number of tour buses in relationship to available enforcement personnel. In an analysis of tour bus spatial dynamics on the National Mall, Daniels *et al.* (2018) found that 37.8% of tour buses violated the idling laws, with negligible enforcement observed. Equally problematic to idling is cruising, where drivers meander through the gateway city's residential areas rather than seeking offsite parking, which is both limited and difficult to access, leading to traffic congestion and increased emissions (Daniels *et al.*, 2018).

As a short-term intervention, the DC Department of Energy & Environment introduced a pilot programme that invites citizens to assist in the improvement of air quality by allowing them to report violations of the engine idling regulations using a DC app designed for this purpose (DC Department of Energy & Environment, 2020a). More significantly, DC's long-term goal to decrease vehicle emissions is summarized in the July 2020 Memorandum of Understanding with 15 states, 'in a commitment to phase out fossil fuel-burning medium- to heavy-duty truck and bus sales by one hundred percent by 2050, with a target for 30 percent of new truck and bus sales to be zero-emission by 2030' (DC Department of Energy & Environment, 2020b).

Active RODs specific to transportation focus on low- to no-emission circulation modes designed to decrease onsite vehicular and tour bus traffic. Three current RODs focus on e-bikes, pedicabs and Segway motorized scooters. E-bikes with motors of less than 750 watts and fully operable pedals are permitted in all areas where traditional bikes are allowed (NPS, 2019). Pedicab companies can seek approval through the National Mall Concessions Department for commercial operation within the National Mall (NPS, 2017b). Segway motorized scooters, while previously forbidden on the National Mall, were approved in 2010 for personal and commercial use, along with other motorized personal mobility devices, with the provision that maximum speed does not exceed 8 mph, and that users must be a minimum of 16 years of age (NPS, 2010c). The ROD specifies designated park sidewalks that are permissible for tour routes (NPS, 2010c). The *Superintendent's Compendium* provides detailed policies regarding transportation and circulation, reiterating regulations found within the RODs and DC laws pertaining to areas such as motorized personal mobility devices, Segway motorized scooters, pedicabs, bicycles, buses, road restrictions and parking guidelines (NPS, 2018a).

12.5 Discussion

The status of DC as an internationally prominent gateway community was evidenced throughout this analysis of PA management plans and practices within the National Mall. While the Green Parks Plan goals were all evidenced in at least one of the three emergent themes of historic landscapes, water and transportation, the park-level translation of these goals cannot be separated from the constraints and challenges specific to DC as a whole. Sustainability guidelines within capital cities must take into account stakeholder initiatives that range from residents' petitions to congressional mandates; as such, managing conflicting perspectives (e.g., Dangi and Gribb, 2018) and adhering to a rulemaking roadmap (e.g., Cook, 2014) become more complicated as the number of visitors, diversity in stakeholders and deferred maintenance costs increase.

Further complicating sustainable tourism initiatives in gateway communities is the fact that capital cities are more vulnerable to political unrest than other tourism destinations (Nickeas, 2021). The most carefully deliberated plans and guidelines may become impracticable in the face of political upheaval. For instance, the National Mall greenspace public use assurances, as outlined within the turf restoration ROD and detailed in advance of the 20 January 2021 US Presidential Inauguration, were suspended following the 6 January 2021 riot at the US Capitol. While the Covid-19 pandemic had already resulted in a reduced-scale onsite approach, to promote social distancing, the focus of the inaugural ceremonies had to pivot to an exclusive emphasis on safety and security, resulting in the non-presence of tourists in the wake of shuttered businesses, the shutdown of all major bridges and public transportation into DC, the closure of the National Mall greenspace for over a week and repeated calls for the public to celebrate remotely rather than venturing into DC (Davies and Jouvenal, 2021). With over 26,000 National Guard troops deployed to DC to ensure a peaceful transfer of power (Altman, 2021), the provisions that ensure First Amendment freedom of expression, as well as any focus on sustainability principles, were unviable following the deadly insurrection, with NPS posting urgent alerts regarding the necessity of the closures in light of the riot (Litterst, 2021).

As the conservators of globally recognizable assets, gateway city administrators must lead by example to ensure a viable quality of life for residents and support a superior experience for visitors while maintaining and protecting cultural and environmental resources (Brash *et al.*, 2011). Stakeholder engagement does not stop at the geographical boundary of a capital city and, instead, extends to the nation as a whole (Daum, 2005). As such, the sustainability steps and missteps are chronicled and widely circulated for inspection and comment. This point was exemplified by the National Mall Plan public comment process, where over 75% of the over 30,000 comments originated from outside the resident population (NPS, 2010a).

12.6 Conclusion

The objective of this chapter was to illuminate the utility and application of the Green Parks Plan goals within an urban context. Engagement and buy-in from diverse urban stakeholders and corporate sponsors, combined with coverage by prestigious media sources, ensure that sustainability issues will be acknowledged, yet the movement from knowledge to action often remains elusive. For instance, the twice-daily flooding of the Tidal Basin area and the gradual sinking of the surrounding area have been recognized for over two decades, yet little progress has been made due to consistent funding shortages (National Trust, 2020). As such, while the National Mall maintains a high-profile status, the NPS management team faces sustainability obstacles that may, at times, seem insurmountable.

By evaluating National Mall policies and projects in light of NPS Green Parks Plan goals, this chapter has highlighted the importance of inclusive sustainable planning. For gateway communities, such as DC, the needs of residents and tourists are inexorably linked and must be considered in tandem and in light of the political climate of the nation as a whole. Greenspace, water and transportation planning are fundamental to all PAs, yet the application of associated strategies will be unique from park to park. This analysis can serve as a model for contextualizing national sustainability guidelines on a local level for urban gateway destinations.

The lens of this chapter has focused almost exclusively on Washington DC, thus highlighting the unique considerations of an internationally prominent capital city. This restricted view can also be seen as the primary limitation of the research, in that the emergent themes should not be viewed as generalisable to other gateway destinations. Nor can it be assumed that the Green Parks Plan goals would apply similarly to other NPS holdings. Future research is suggested to address this shortcoming, with comparative studies recommended at the gateway city level and, separately, through a comparison of urban parks.

By evaluating specific NPS policies and projects, this chapter has highlighted the importance of inclusive planning, specifically to include tourist attractions and tourism agencies, often neglected within urban contexts (Moyle *et al.*, 2014). The case study approach illustrates that gateway city planning can be positively influenced when prioritizing iconic attractions, especially in a destination where tourists outnumber local residents. The spill-over benefits of the Green Parks Plan have been, and will continue to be, consequential to the residents of DC and will impact national and international perceptions of this gateway destination.

References

Altman, H. (2021) 26,000 National Guard Troops Came to DC and Protected the Inauguration Without Incident. Now the Drawdown Begins. *Military Times*. Available at: https://www.militarytimes.com/news/your-military/2021/01/21/26000-national-guard-troops-came-to-dc-to-protect-the-inauguration-now-the-drawdown-begins/

Bellini, N., Go, F.M. and Pasquinelli, C. (2017) Urban tourism and city development: Notes for an integrated policy agenda. In: Bellini, N. and Pasquinelli, C. (eds) *Tourism in the City: Towards an Integrative Agenda on Urban Tourism*. Springer International Publishing, Cham, Switzerland, pp. 333–339.

Brash, A., Hand, J. and Orff, K. (2011) *Gateway: Visions for an Urban National Park*. Princeton Architectural Press, Princeton, NJ, USA.

Brown, G.A. (2009) Document analysis as a qualitative research method. *Qualitative Research Journal* 9(2), 27–40.

Childers, D.L., Pickett, S.T., Grove, J.M., Ogden, L. and Whitmer, A. (2014) Advancing urban sustainability theory and action: challenges and opportunities. *Landscape and Urban Planning* 125, 320–328. DOI: 10.1016/j.landurbplan.2014.01.022.

Cook, J.J. (2014) Are we there yet? A roadmap to understanding National Park Service rulemaking. *Society & Natural Resources* 27(12), 1257–1270.

Dangi, T.B. and Gribb, W.J. (2018) Sustainable ecotourism management and visitor experiences: managing conflicting perspectives in Rocky Mountain National Park, USA. *Journal of Ecotourism* 17(3), 338–358.

Daniels, M.J. (2020) Sustainable greenspace management: The National Cherry Blossom Festival. In: Slocum, S.L., Aidoo, A. and McMahon, K. (eds) *The Business of Sustainable Tourism Development and Management*. Routledge, Milton Park, UK, pp. 128–131.

Daniels, M.J., Harmon, L.K., Barrow, M., Park, M., Brayley, R.E. *et al.* (2015) Heritage tourism within a capital city: symbolism and dissonance. *Tourism Review International* 19, 1–18.

Daniels, M.J., Harmon, L.K., Vese Jr., R., Park, M. and Brayley, R. (2018) Spatial optimization of tour bus transport within urban destinations. *Tourism Management* 64, 129–141.

Daum, A. (2005) Capitals in modern history: Inventing urban space for the nation. In: Daum, A.W. and Mauch, C. (eds) *Berlin, Washington, 1800-2000: Capital Cities, Cultural Representation and National Identity*. Cambridge University Press, New York, NY, USA, pp. 3–28.

Davies, E. and Jouvenal, J. (2021, January 14) Inauguration Day Closures Expected to Include the Mall, Major Bridges. *The Washington Post*. Available at: https://www.washingtonpost.com/local/national-mall-closed-inaugruation-day/2021/01/14/1ca4e540-561c-11eb-a817-e5e7f8a406d6_story.html

Daymon, C. and Holloway, I. (2002) *Qualitative Research Methods in Public Relations and Marketing Communications*. Routledge, London, UK.

DC Department of Energy & Environment (2020a) *Engine Idling Citizen Enforcement Program*. Available at: https://doee.dc.gov/service/engine-anti-idling-law (accessed 2 December 2020).

DC Department of Energy & Environment (2020b) *Medium and Heavy-duty Zero Emission Vehicles*. Available at: https://doee.dc.gov/service/mhd-ze-vehicles (accessed 2 December 2020).

DC Department of Energy & Environment (2020c) *Turn It Off*. Available at: https://doee.dc.gov/sites/default/files/dc/sites/ddoe/service_content/attachments/turn-it-off-combined.pdf (accessed 2 December 2020).

DC.Gov (2020) *Washington, DC: New Population, New Year, New Housing*. Available at: https://dc.gov/release/new-population-new-year-new-housing (accessed 18 November 2020).

Eagles, P.F.J. (2014) Research priorities in park tourism. *Journal of Sustainable Tourism* 22(4), 528–549.

Fava, N. and Rubio, S.P. (2017) From Barcelona: the pearl of the Mediterranean to bye bye Barcelona. In: Bellini, N. and Pasquinelli, C. (eds) *Tourism in the city*. Springer, Cham, Switzerland, pp. 285–295.

Frauman, E. and Banks, S. (2011) Gateway community resident perceptions of tourism development: incorporating importance-performance analysis into a limits of acceptable change framework. *Tourism Management* 32, 128–140.

Fuller, T.K. (2016) Mapping environmental justice: a framework for understanding sustainability at the neighborhood scale in Indianapolis. In: Gatrell, J.D., Jensen, R.R., Patterson, M.W. and Hoalst-Pullen, N. (eds) *Urban Sustainability: Policy and Praxis*. Springer, Cham, Switzerland, pp. 201–215.

Hardy, A. (2005) Using grounded theory to explore stakeholder perceptions of tourism. *Journal of Tourism and Cultural Change* 3(2), 108–133.

Harmon, L.K., Daniels, M.J., Brayley, R.E. and Park, M. (2011) Is a nation's front yard becoming a dumping ground? A critical analysis of greenspace within the United States National Mall & Memorial Parks. *Journal of Tourism and Leisure Studies* 2, 269–292.

Kavaratzis, M. (2017) The participatory place branding process for tourism: linking visitors and residents through the city brand. In: Bellini, N. and Pasquinelli, C. (eds) *Tourism in the City: An Integrated Approach*. Springer, Cham, Switzerland, pp. 93–107.

Kerr, W.R. (2003) *Tourism Public Policy, and the Strategic Management of Failure*. Pergamon, Oxford, UK.

Litterst, M. (2021) *Washington Monument Closed for COVID-19 Prevention*. National Park Service. Available at: https://www.nps.gov/nama/learn/news/wamoclosed.htm (accessed 1 February 2021).

Maxim, C. (2015) Drivers of success in implementing sustainable tourism policies in urban areas. *Tourism Planning & Development* 12(1), 37–47.

Maxim, C. (2016) Sustainable tourism implementation in urban areas: a case study of London. *Journal of Sustainable Tourism* 24(7), 971–989.

Moyle, B.D., McLennan, C.J., Ruhanen, L. and Weiler, B. (2014) Tracking the concept of sustainability in Australian tourism policy and planning documents. *Journal of Sustainable Tourism* 22(7), 1037–1051.

Muangasame, K. and McKercher, B. (2015) The challenge of implementing sustainable tourism policy: A 360-degree assessment of Thailand's '7 Greens sustainable tourism policy'. *Journal of Sustainable Tourism* 23(4), 497–516.

National Park Service (2007) *Best Management Practices Used at Urban Parks in National and International Locations; A Background Report for the National Mall Plan*. Available at: https://www.nps.gov/nationalmallplan/Documents/National%20Mall%20Plan%20-%20Best%20Management%20Practices%20Used%20at%20Urban%20Parks%20in%20National%20and%20International%20Locations.pdf (accessed 17 November 2020).

National Park Service (2010a) *National Mall and Memorial Parks Climate Action Plan*. Available at: https://www.nps.gov/subjects/climatechange/upload/NAMA-Final-Climate-Action-Plan.pdf (accessed 8 November 2020).

National Park Service (2010b) *National Mall Plan*. Available at: https://www.nps.gov/nationalmallplan/FEISdocs.html (accessed 17 November 2020).

National Park Service (2010c) *Revised Record of Determination for a Public Use Limitation on Segway and Personal Mobility Vehicles*. Available at: https://www.nps.gov/nama/learn/management/record-of-determination-for-a-public-use-limitation-on-segway-and-personal-mobility-vehicles.htm (accessed 2 December 2020).

National Park Service (2016) *Green Parks Plan: Advancing Our Mission through Sustainable Operations*. Available at: https://www.nps.gov/subjects/sustainability/upload/NPS-Green-Parks-Plan-2016.pdf (accessed 20 November 2020).

National Park Service (2017a) *National Mall and Memorial Parks: History & Culture*. Available at: https://www.nps.gov/nama/learn/historyculture/index.htm (accessed 2 December 2020).

National Park Service (2017b) *Pedicab Commercial Use Application*. Available at: https://www.nps.gov/nama/learn/management/upload/2017-PEDICAB-CUA-Application-with-Instructions-NPS-Form-10-550.pdf (accessed 2 December 2020).

National Park Service (2018a) *Superintendent's Compendium*. Available at: https://www.nps.gov/nama/learn/management/superintendent-s-compendium.htm (accessed 21 November 2020).

National Park Service (2018b) *Tidal Basin. Washington, DC*. Available at: https://www.nps.gov/articles/dctidalbasin.htm (accessed 20 February 2021).

National Park Service (2019) *Record of Determination for Allowing E-Bikes*. Available at: https://www.nps.gov/nama/learn/management/ebikes-rod.htm (accessed 2 December 2020).

National Park Service (2020a) *National Mall and Memorial Parks: How Many People Visit the Park?*. Available at: https://www.nps.gov/nama/faqs.htm (accessed 22 December 2020).

National Park Service (2020b) *Record of Determination for a Partial and Temporary Closure of a Portion of the National Mall for Turf Restoration*. Available at: https://www.nps.gov/nama/learn/management/turfrod.htm (accessed 1 December 2020).

National Trust for Historic Preservation (2019) *National Mall Tidal Basin Ideas Lab Manual*. Available at: https://nthp-savingplaces.s3.amazonaws.com/2019/10/25/10/16/01/413/TidalBasin_IdeasLab_View_20191024.pdf (accessed 20 February 2021).

National Trust for Historic Preservation (2020) *Tidal Basin Ideas Lab*. Available at: https://tidalbasinideaslab.org/ (accessed 1 December 2020).

Nickeas, P. (2021) *"That's like putting gasoline on a fire." Some states are more vulnerable to unrest than others*. CNN, January 15. Available at: https://www.cnn.com/2021/01/15/us/capitol-insurrection-threat-states/index.html (accessed 31 January 2021).

Pearce, D.G. (2015) Urban management, destination management and urban destination management: a comparative review with issues and examples from New Zealand. *International Journal of Tourism Cities* 1(1), 1–17.

Pew (2020) *National Park Deferred Maintenance Needs*. Available at: https://www.pewtrusts.org/en/research-and-analysis/data-visualizations/2018/national-park-deferred-maintenance-needs (accessed 23 November 2020).

Repanshek, K. (2019) Human resource problems straining workforce in some national parks. *National Parks Traveler*. Available at: https://www.nationalparkstraveler.org/2019/08/human-resource-problems-straining-workforce-some-national-parks (accessed 21 November 2020).

Slocum, S.L. (2017) Operationalising both sustainability and neo-liberalism in protected areas: Implications from the USA's National Park Service's evolving experiences and challenges. *Journal of Sustainable Tourism* 25(12), 1848–1864.

Stankey, G.H., Cole, D.N., Lucas, R.C., Petersen, M.E. and Frissell, S.S. (1985) *The limits of acceptable change (LAC) system for wilderness planning (General technical report INT-176)*. USDA Forest Service, Intermountain Forest and Range Experiment Station, Ogden, UT, USA.

Wang, C.Y. and Miko, P.S. (1997) Environmental impacts of tourism on US national parks. *Journal of Travel Research* 35(4), 31–36.

Zheng-Xin, W. and Pei, L. (2014) A systems thinking-based grey model for sustainability evaluation of urban tourism. *Kybernetes* 43(3), 462–479.

13 Conclusion

Susan L. Slocum[1]* and John B. Read IV[2]
[1]George Mason University, Virginia, USA; [2]University of Waterloo, Canada

Abstract

This chapter provides an overview of *Tourism Transformations in Protected Area Gateway Communities* by extracting common themes that run throughout the book. It then situates these themes within the larger scholarly work on gateway communities bordering protected areas (PAs). The common themes include: impactful projects; expanding the stakeholder base; environment and social; and best practice and theoretical frameworks.

Keywords: gateway communities, protected area (PA) management, community engagement, collaboration/partnerships

13.1 The Interrelationship of Protected Areas and Gateway Communities

Tourism Transformations in Protected Area Gateway Communities is a first step in acknowledging that gateway communities, those that serve as an entry and exit point to protected areas (PAs) (Frauman and Banks, 2011), face unique challenges and opportunities related to tourism visitation. Burghardt (1971) was the first to acknowledge the distinctive characteristics of gateway communities as those that share a common boundary with PAs, where land management ends on one side of this arbitrary, permeable border and heterogeneous society exists on the other side. For centuries, these communities were viewed as barriers to conservation, often resulting in the displacement of peoples (Chatty and Colchester, 2002) and/or the restriction of traditional livelihoods and cultural practices (Nyaupane and Poudel, 2011). Recent scholarly work (Aguirre, 2006; Frauman and Banks, 2011; Joyner *et al.*, 2019) has emphasized the need for a more inclusive form of governance for PAs as a means to enhance community support for conservation efforts and improve the quality of life for gateway residents. Both healthy natural and cultural areas, as well as vigorous gateway communities, reinforce tourism visitation, which, in turn, increases revenue for both parties and furthers government and civil society's patronage; all of which ensures the survival of PAs, the gateway communities and the services they both provide. Briassoulis (2002) articulates the interdependence of these three variables when she writes, 'economic wellbeing declines, environmental conditions worsen, social injustice grows, and tourist satisfaction drops' when natural resources are overused as a result of tourism activities (p. 1065).

This book has highlighted many of the situations that have arisen from a recent change in the philosophical and practical approach to PA management. Administrators have been forced to come to terms with the reality that PAs are not

*Corresponding author: sslocum@gmu.edu

 Tourism Transformations in Protected Area Gateway Communities (eds S.L. Slocum *et al.*)
DOI: 10.1079/9781789249033.0013

islands among themselves, but part of a larger landscape and society that requires collaboration and partnership to enhance conservation efforts, community relationships and visitor satisfaction. As an example, invasive species from outside PAs can have a devastating effect on protected ecosystems, and so can extensive development (Slocum, 2016), whereas wildlife may leave protected areas, causing damage to agriculture and harming local residents (Sindiga, 1995). As another example, gateway communities in a state of economic or social decline can blight the landscape as tourism arrives at PAs, reducing visitation and/or tourist satisfaction (Poudyal *et al.*, 2013). Forsaking one to spite the other is counter-intuitive; the strength of conservation lies in long-term strategies that ensure gateway communities are partners, not adversaries, and that communities maintain their vitality, so they no longer need to exploit natural resources (Jones *et al.*, 2015).

Moreover, gateway communities *and* PAs jointly form the tourism destination, requiring a more holistic approach to long-term planning and impact mitigation. As an example, Supak *et al.* (2015, p. 174) explain,

> these gateway communities are often seen not only as the portals to cherished landscapes, but as the purveyors of food, lodging, transportation and other business support for visitors ... while the sites attract millions of visitors because of their scenic beauty and outdoor recreation opportunities, only a small percentage of these visitors can be accommodated within the most popular parks each night.

Destination marketing and management literature highlights the need for a common vision to inform a communication strategy that encompasses authentic placemaking narratives, which are acceptable, not only to major attractions, but to residents and visitors alike (Hultman and Hall, 2012). Sense of place instils long-term concern for resources and may encourage both repeat visitation and philanthropic motivation that leads to financial and volunteer support for PAs (Farnum *et al.*, 2005). Communities may have an identity that extends beyond the PA, to include historic, cultural or diversity characteristics that impact sense of place and should influence marketing messages.

As stated by Wiltshier and Read in the Introduction, the goal of this book is to 'provide mutually beneficial alternatives to PA–community relationships that support both the sustainability of the PAs themselves, as well as community wellbeing'. This has been done through a presentation of multiple case studies that bring insightful analysis to unique partnerships and collaborative efforts in a wide-ranging array of destinations. These emerging alliances take many forms and are as idiosyncratic as the parks and communities that form them. While this book is not exhaustive in its examples, we hope it provides concrete illustrations that stimulate creative juices towards potential collaborations, a widening of potential projects and a more inclusive form of protected area governance. This book offers a number of thematic commonalities that unite the varied chapters and presents a fuller picture of gateway communities, specifically, and PA management, generally. These themes include: impactful projects; expanding the stakeholder base; environment and social interactions; and best practice and theoretical frameworks.

13.2 Impactful Projects

One of the most important aspects of this book is the clarity with which problems are defined and solutions found that address both the needs of PAs and the communities mainly impacted by visitation. Literature shows that many of the tourism impacts are felt most strongly within gateway communities (Bergstrom and Harrington, 2018; Lackey and Bricker, 2020) and that the magnitude of impacts can overwhelm both communities and protected areas (McMahon, 1999; Kurtz, 2010; Slocum, 2016). Therefore, finding manageable, impactful projects that require collaboration between all stakeholders is often a first step in developing effective partnerships. Kline *et al.* (2012) attribute success to 'selecting a small, visible, and uncontroversial project as a first community endeavor' (n.p.). New collaborations also require initial successes to maintain enthusiasm for newly established collaborations (Slocum and Kline, 2014).

Chapter 6 and Chapter 9 utilize the United Nations' Sustainable Development Goals as a means to address not only the most pressing needs of the community, but to find projects that encourage meaningful and successful mitigations that are viable, visible and impactful. In Chapter 6, González-Herrera *et al.* show how park concessionaires send compostable waste to local farms and then buy back agricultural products grown with these waste materials, along with sustainable landfill supported by corporate entities, local charity groups, park stakeholders and local businesses. In Chapter 9, Rodrigues *et al.* acknowledge that the focus on water sanitation projects has resulted in an increase in 'belonging and identity in the community through the recognition of the health benefits and improved quality of life for residents'. Both of these examples have shown measurable impacts that improve overall quality of life for the residents and the visitor experience.

In Chapter 12, Daniels and Slocum showcase the National Mall's Green Parks Plan, which breaks down historic landscapes, water mitigation and transportation issues as priorities for the Washington DC area. In doing so, local stakeholders have been able to instil a number of manageable projects that have directly influenced the sustainability of both the park and the greater metropolitan area. Their chapter demonstrates that 'gateway city planning can be positively influenced when prioritizing iconic attractions, especially in a destination where tourists outnumber local residents. The spillover benefits of the Green Parks Plan have been, and will continue to be, consequential to the residents of DC'.

13.3 Expanding the Stakeholder Base

Considered to be a touchstone of tourism planning and development, involving stakeholders is critical. While Pomeroy and Douvere (2008) advance this through the stakeholder approach, others like Stewart and Draper (2007) discuss how understanding stakeholder perceptions can be influential in recognizing the variation between groups. Early work by Holden (2003) called for the inclusion of a new environmental ethic that emphasized a non-anthropogenic perspective, while more recently, Read IV and Grimwood (2021) continue to advocate for a more central role of ethics in ecotourism development, especially when working with cross-cultural stakeholders, by adopting the perspective of culturally sensitive universalism. Moreover, as visits to outdoor recreation areas increase in response to the need for COVID safe vacation options, Roxas *et al.* (2020) note that stakeholder involvement will be critical in rebooting tourism destinations. Demonstrably, there is a need for continued stakeholder involvement which requires both innovative, as well as nuanced, approaches to achieve buy-in in an increasingly global and post-pandemic future.

Invariably, every chapter in this book addresses stakeholders due to its emphasis on PAs and gateway communities. Importantly, however, the book also highlights some of the innovative and nuanced approaches that epitomize the central theme of expanding the stakeholder base. As discussed in Chapter 4 and Chapter 10, stakeholder approaches can build fruitful relationships in existing PAs that achieve sustainability goals as well as allow for creativity that can accelerate certain bureaucratic processes. In Chapter 4, Lackey *et al.* deliver a shared vision based on environmental ethics that has led to necessary partnerships and the development of social capital. Sleipness *et al.*, in Chapter 10, write 'where natural features, including topography, hydrological features and sensitive ecosystems limit the range of potential activities suited for the site, partnerships play an important role in allowing planning and design to address these challenges at an adequate scale'.

Moscardo, in Chapter 11, offers further innovation by advancing the concept of storying gateway communities and their associated PAs through the involvement of communities as stakeholders in influencing the stories that are communicated to tourists. Alternatively, Chapter 8 and Chapter 9 demonstrate how stakeholder consultation for existing PAs can illuminate the tensions and conflicts that exist or, as seen in Chapter 2, the manner in which communities fluctuate and take ownership of the PAs near which they reside. Within burgeoning PAs, stakeholder groups in a community can serve as critical advocates, as seen in Chapter 5.

Ventura *et al.*, in Chapter 7, highlight the importance of environmental non-governmental organizations (ENGOs) in the stakeholder mix, specifically ENGOs that possess expertise in both PA management and community relations. They found that the level of involvement of these groups had a direct impact on access to capital by gateway communities within tourism destinations. They write, 'ENGOs are not only part of these island gateway communities but give them an opportunity to participate actively in local tourism management'.

13.4 Environment and Social Interactions

Always intertwined, PAs and the flora and/or fauna that are located within these areas of 'wilderness' can often be impossible to tease apart. Humans equally play a role in PAs, and the link between local communities and conservation efforts has continued to grow (Nepal, 1997). However, as mentioned by Wiltshier and Read in the introduction, PAs have often been situated on lands where Indigenous peoples have been displaced and/or disallowed access (Chatty and Colchester, 2002). This may create complicated relationships between the environmental and social worlds, which can often be conveyed through the place attachment of individuals to a location (Altman and Low, 1992) or through official or unofficial policies that craft the identity of a place (Dredge and Jenkins, 2003).

The relationship between the environment and social worlds features prominently in Chapter 3, Chapter 5 and Chapter 8. In Chapter 3, Bohn and Hall discuss the evolution of Lapland's identity post Second World War and how political and social forces, through a political economy, have influenced the presentation of the environment. In turn, the community has developed an identity that is aligned with the arctic environment and has crafted tourism opportunities to align with Lapland as the gateway to the Arctic. Chapter 5 provides a differing example as Walker and Lawrence discuss a marine management area in the Caribbean and how protection of the environment is reported by residents as something they are invested in, as they see increased livelihood opportunities, such as employment as ecotourism guides. Additionally, they highlight how the gateway community advocates to protect the marine management area. The community acts as a stakeholder that encourages responsible management so that livelihoods can be maintained. Alternatively, in Chapter 8, Dela Santa *et al.* demonstrate that protected areas can create conflict between environmental and social dynamics. Through exploring the relationship of the Ivatans – an Indigenous people in the Philippines – the chapter demonstrates the difficulties that can occur when PAs are created and regulated without considering the needs of the people whose identities are tied to these locations. Importantly, Chapter 8 highlights the dangers of protected area management where the Ivatans are limited in their ability to practise traditional heritage activities, reminiscent of what is mentioned by Chatty and Colchester (2002).

13.5 Best Practice and Theoretical Frameworks

Clearly, the best relationships between PAs and their gateway communities are built on robust stakeholder involvement as seen within all the chapters of this book. As advocated by Grimwood and Doubleday (2013) regarding Indigenous peoples and PAs, adaptive co-management offers ethical opportunities for those from differing cultural backgrounds and worldviews to embrace governance as the co-mingling of perspectives, approaches and knowledge. An alternative stakeholder approach specifically discussed in Chapter 8 suggests opportunities for co-management of PAs by their gateway communities with the intent to 'minimize, if not prevent, the challenges and development pressures'. Stakeholder engagement in some locations can also require a fundamental shift in how PAs are perceived, as highlighted in Chapter 4 by Lackey *et al.* who note that management perspectives in some US National Parks must shift from the mentality of the PA as an island that is isolated from the surrounding communities to one as a partner. Fortunately, a national scenic trail – the Appalachian Trail – and Zion National Park offer models which demonstrate how gateway communities, national parks and other stakeholders can integrate and share different aspects of PA management.

We can also see theoretical frameworks within the chapters that illuminate new ways of understanding the relationships between PAs and their gateway communities. In Chapter 8, we see the incorporation of discourse analysis, specifically a Foucaultian approach (Akdağ and Swanson, 2018), to illuminate the manner in which power shapes relationships and underlying discourses. While in Chapter 11, we see how Moscardo incorporates storying frameworks as a story-turn within tourism (Moscardo, 2020) to demonstrate the role of community members, PA managers, tourism planners and developers as stakeholders who can generate and convey the stories associated with both the gateway community and their PA. Alternatively, Bohn and Hall in Chapter 3 demonstrate how approaching tourism in Lapland from a 'geographical political economy approach to state rescaling, informed by regulation theory' focused on 'how the relationships of public, private and third sectors under capitalism are organized into a regime of accumulation and social regulation in space and time' (Sheppard, 2016). They use this to showcase the manner in which geopolitical and economic drivers have reshaped the identity and perception of Lapland from the traditional territory of the Indigenous Sami people to an area serving as gateway to the Arctic.

References

Aguirre, J.A. (2006) Linking national parks with its gateway communities for tourism development in central America: Nindiri, Nicaragua, Bagazit, Costa Rica and Portobelo, Panama. *Pasos. Revista de Turismo y Patrimonio Cultural* 4(3). DOI: 10.25145/j.pasos.2006.04.024.

Akdağ, E. and Swanson, D.M. (2018) Ethics, power, internationalisation and the postcolonial: a Foucauldian discourse analysis of policy documents in two Scottish universities. *European Journal of Higher Education* 8(1), 67–82. DOI: 10.1080/21568235.2017.1388188.

Altman, I. and Low, S.M. (1992) *Place Attachment. Human Behaviour and Environment: Advances in Theory and Research*. Plenum Press, New York, NY, USA.

Bergstrom, R.D. and Harrington, L.M. (2018) Understanding agents of change in amenity gateways of the Greater Yellowstone region. *Community Development* 49(2), 145–160. DOI: 10.1080/15575330.2017.1416648.

Briassoulis, H. (2002) Sustainable tourism and the question of the commons. *Annals of Tourism Research* 29(4), 1065–1085. DOI: 10.1016/S0160-7383(02)00021-X.

Burghardt, A.F. (1971) A hypothesis about gateway cities. *Annals of the Association of American Geographers* 61(2), 269–285. DOI: 10.1111/j.1467-8306.1971.tb00782.x.

Chatty, D. and Colchester, M. (2002) *Conservation and Mobile Indigenous Peoples: Displacement, Forced Settlement, and Sustainable Development*. Berghahn Books, New York, NY, USA.

Dredge, D. and Jenkins, J. (2003) Destination place identity and regional tourism policy. *Tourism Geographies* 5(4), 383–407. DOI: 10.1080/1461668032000129137.

Farnum, J., Hall, T. and Kruger, L.E. (2005) Sense of Place in Natural Resource Recreation and Tourism: an Evaluation and Assessment of Research Findings. Gen. Tech. Rep. PNW-GTR-660. U.S. Department of Agriculture, Forest Service, Pacific Northwest Research Station, Portland, OR, USA.

Frauman, E. and Banks, S. (2011) Gateway community resident perceptions of tourism development: incorporating importance-performance analysis into a limits of acceptable change framework. *Tourism Management* 32(1), 128–140. DOI: 10.1016/j.tourman.2010.01.013.

Grimwood, B.S.R. and Doubleday, N.C. (2013) From river trails to adaptive CO- management: learning and relating with inuit inhabitants of the thelon river, canada. *Indigenous Policy Journal* 23(4), Article. Available at: http://indigenouspolicy.org/index.php/ipj/article/view/146 (accessed 7 February 2022).

Holden, A. (2003) In need of new environmental ethics for tourism? *Annals of Tourism Research* 30(1), 94–108. DOI: 10.1016/S0160-7383(02)00030-0.

Hultman, J. and Hall, C.M. (2012) Tourism place-making: governance of locality in Sweden. *Annals of Tourism Research* 39(2), 547–570.

Jones, B.T., Diggle, R.W. and Thouless, C. (2015) From exploitation to ownership: wildlife-based tourism and communal area conservancies in Namibia. In: Van der Duim, R., Lamers, M., and Van Wijk, J.

(eds) *Institutional Arrangements for Conservation, Development and Tourism in Eastern and Southern Africa*. Springer, Dordrecht, The Netherlands, pp. 17–37.

Joyner, L., Lackey, N.Q. and Bricker, K.S. (2019) Community engagement: an appreciative inquiry case study with Theodore Roosevelt National Park gateway communities. *Sustainability* 11(24), 7147. DOI: 10.3390/su11247147.

Kline, C., Brown-Bochicchio, C. and Beedle, J. (2012) *A report on the impacts of the Bakersville improvement group and HandMade in America*. Recreation and Leisure Studies/Center for Sustainable Tourism, East Carolina University, Greenville, NC, USA.

Kurtz, R.S. (2010) Public lands policy and economic trends in gateway communities. *Review of Policy Research* 27(1), 77–88. DOI: 10.1111/j.1541-1338.2009.00428.x.

Lackey, N.Q. and Bricker, K. (2020) The relationship between parks and protected areas and gateway communities in the United States and Canada: a scoping review. *Journal of Park and Recreation Administration*. DOI: 10.18666/JPRA-2020-10390.

McMahon, E.T. (1999) Gateway communities. *Planning Commissioners Journal* 34, 6–7.

Moscardo, G. (2020) The story turn in tourism: forces and futures. *Journal of Tourism Futures* 7(2), 168–173. DOI: 10.1108/JTF-11-2019-0131.

Nepal, S.K. (1997) Sustainable tourism, protected areas and livelihood needs of local communities in developing countries. *International Journal of Sustainable Development & World Ecology* 4(2), 123–135. DOI: 10.1080/13504509709469948.

Nyaupane, G.P. and Poudel, S. (2011) Linkages among biodiversity, livelihood, and tourism. *Annals of Tourism Research* 38(4), 1344–1366. DOI: 10.1016/j.annals.2011.03.006.

Pomeroy, R. and Douvere, F. (2008) The engagement of stakeholders in the marine spatial planning process. *Marine Policy* 32(5), 816–822. DOI: 10.1016/j.marpol.2008.03.017.

Poudyal, N.C., Paudel, B. and Tarrant, M.A. (2013) A time series analysis of the impact of recession on national park visitation in the United States. *Tourism Management* 35(4), 181–189. DOI: 10.1016/j.tourman.2012.07.001.

Read IV, J.B. and Grimwood, B.S.R. (2021) Ecotourism development through culturally sensitive universalism. In: Fennell, D.A. (ed.) *Routledge Handbook of Ecotourism*. Routledge, London, UK, pp. 103–116.

Roxas, F.M.Y., Rivera, J.P.R. and Gutierrez, E.L.M. (2020) Mapping stakeholders' roles in governing sustainable tourism destinations. *Journal of Hospitality and Tourism Management* 45(16), 387–398. DOI: 10.1016/j.jhtm.2020.09.005.

Sheppard, E. (2016) *Limits to globalization. Disruptive geographies of capitalist development*. Oxford University Press, Oxford, UK.

Sindiga, I. (1995) Wildlife based tourism in Kenya: land use conflicts and government compensation policies over protected areas. *Journal of Tourism Studies* 6(2), 45–55.

Slocum, S.L. (2016) Operationalising both sustainability and neo-liberalism in protected areas: implications from the USA's National Park Service's evolving experiences and challenges. *Journal of Sustainable Tourism* 25(12), 1848–1864. DOI: 10.1080/09669582.2016.1260574.

Slocum, S. and Kline, C. (2014) Regional resilience: opportunities, challenges and policy messages from Western North Carolina. *Anatolia* 25(3), 403–416. DOI: 10.1080/13032917.2014.888673.

Stewart, E.J. and Draper, D. (2007) A collaborative approach to understanding local stakeholder perceptions of tourism in Churchill, Manitoba (Canada). *Polar Geography* 30(1–2), 7–35. DOI: 10.1080/10889370701666689.

Supak, S., Brothers, G., Bohnenstiehl, D. and Devine, H. (2015) Geospatial analytics for federally managed tourism destinations and their demand markets. *Journal of Destination Marketing & Management* 4(3), 173–186. DOI: 10.1016/j.jdmm.2015.05.002.

Index

Note:- Page numbers in **bold** type refer to figures and page numbers in *italic* type refer to tables.